Lecture Notes
in Control and Information Sciences 252

Editor: M. Thoma

Springer
London
Berlin
Heidelberg
New York
Barcelona
Hong Kong
Milan
Paris
Santa Clara
Singapore
Tokyo

Murti V. Salapaka and Mohammed Dahleh

Multiple Objective Control Synthesis

With 17 Figures

Springer

Authors
Murti V. Salapaka, PhD
Department of Electrical Engineering, Iowa State University, Ames, Iowa 50011, USA

Mohammed Dahleh, PhD
Department of Mechanical Engineering, University of California, Santa Barbara, CA 93106, USA

ISBN 1-85233-256-5 Springer-Verlag London Berlin Heidelberg

British Library Cataloguing in Publication Data
Salapaka, Murti V.
Multiple objective control synthesis. - (Lecture notes in control and information sciences ; 252)
1.Automatic control - Mathematical models
I.Title II.Dahleh, Mohammed
629.8'312
ISBN 1852332565

Library of Congress Cataloging-in-Publication Data
A catalog record for this book is available from the Library of Congress

Typesetting: Camera ready by authors
Printed and bound at the Athenæum Press Ltd., Gateshead, Tyne & Wear
69/3830-543210 Printed on acid-free paper SPIN 10746705

To my parents: Shri. S. Prasada Rao and Smt. S. Subhadra
(Murti V. Salapaka)

To my wife: Marie Dahleh
(Mohammed Dahleh)

Preface

Many control design tasks which arise from engineering objectives can be effectively addressed by an equivalent convex optimization problem. The significance of this step stems from efficient computational tools that are available for solving such problems. However, in most cases the resulting convex optimization problem is infinite dimensional. Thus effective finite dimensional convex approximations are needed to complete the control design task.

Researchers have employed advanced mathematical tools to exploit and expose the structure of the resulting convex optimization problem with the objective of obtaining computable ways of obtaining the controller. One of the striking insights obtained by such tools was that certain optimal control problems are equivalent to finite dimensional programming problems. Thus seemingly infinite dimensional problems can be converted to finite dimensional problems.

Even when the convex optimization problem is truly infinite dimensional or when a finite dimensional characterization is not established, the philosophy that has emerged is to establish finite dimensional approximations to the infinite dimensional problem, which guarantee the optimal performance within any prespecified tolerance. Here too researchers have borrowed advanced mathematical tools to exploit the underlying structure of the problem.

One of the difficulties faced by a researcher in this area is the lack of a source where a comprehensive treatment of the tools employed is given. In this book we attempt to fill this gap by developing various topological and functional analytic tools that are commonly used in the formulation and solution of a class of optimal control problems.

Efficient design techniques for multi-objective controllers are necessary because often a single measure fails to capture the design performance objective. The standard $\mathcal{H}_2, \mathcal{H}_\infty$ and ℓ_1 designs are incapable of handling such multi-objective concerns because they optimize a single measure which is no guarantee of performance with respect to some other measure.

An important subclass of multi-objective problems is the class of problems for synthesizing optimal controllers which guarantee performance with respect to both the $\mathcal{H}_2$ measure and relevant time domain measures. The $\mathcal{H}_2/\ell_1$ problem is an example which falls in this class of multi-objective problems where the objective is the design of controllers which optimally reject white noise while guaranteeing stability margins with respect to uncertainty. We apply the developed mathematical tools to such problems where the $\mathcal{H}_2$ measure and time domain measures on the performance of the closed loop system can be incorporated in a natural manner.

Organization of the Book

This book can be divided into two parts. The first part constitutes Chapters 1, 2 and 3 where the mathematical machinery is developed. In the second part, (Chapters 4, 5, 6, 7, 8 and 9) various control design problems are formulated and solved.

In Chapter 1 we introduce the basic topological concepts. The importance of continuity and compactness with regards to optimization is established. We take a top-down approach where the spaces described have more and more structure as one proceeds through the chapter.

In Chapter 2 functions on vector spaces and weak topologies are studied. Motivation for why weak topologies are important is elucidated. The Banach-Alaoglu result on compactness of bounded sets in weak topologies is proven. Important results on sublinear functions are given which prove to be instrumental in studying convex sets and functions.

The chapter on convex analysis is the culmination of the mathematical treatment given where we establish the Kuhn-Tucker-Lagrange duality result.

Chapter 4 develops a paradigm where control design objectives can be stated precisely and in an effective manner. Youla parameterization of all closed-loop maps achievable via stabilizing controllers is developed.

Chapters 5 and 6 study single-input single-output systems. In Chapter 5 the ℓ_1 norm of the closed loop is minimized while keeping its $\mathcal{H}_2$ norm below a prespecified value. In Chapter 6 a weighted combination of the $\mathcal{H}_2$ norm of the closed loop and various other relevant time domain measures is minimized over all stabilizing controllers. Exact solutions to the problems formulated are given and continuity of the solutions with respect to change in parameters is established. Even though these problems address single-input single-output systems, they serve to highlight the nature of mixed objective problems involving the two norm of the closed-loop and time domain measures.

In Chapter 7 the square case of the of the $\mathcal{H}_2 - \ell_1$ problem is studied. It is shown that the problem is equivalent to a single finite dimensional quadratic programming problem.

In Chapter 8 the interplay of the $\mathcal{H}_2$ and the ℓ_1 norms of the closed-loop in the general multiple-input multiple-output setting is studied. It is shown that controllers can be designed to achieve performance within any given tolerance of the optimal performance via finite dimensional quadratic programming. The design methodology avoids many problems associated with zero-interpolation based methods.

Chapter 9 tackles a non-convex problem where the $\mathcal{H}_2$ norm of the closed loop is minimized while guaranteeing a specified level of ℓ_1 performance for a collection of plants in a certain class. It is shown that this robust performance problem can be solved via a simplex-like procedure.

Acknowledgements

We would like to thank the students at University of California at Santa Barbara who made valuable suggestions at various stages of the book. In particular we would like to thank Srinivasa Salapaka who proofread the first four chapters of the book. We would like to acknowledge Petar Kokotovic for the encouragement he provided in publishing this book.

The methodology on multiple objective problems in the book was largely shaped in collboration with Petros Voulgaris. The results presented in Chapter 8 were obtained in collaboration with Mustafa Khammash. The results presented in Chapter 9 were obtained in collaboration with Antonio Vicino and Alberto Tesi.

We would like to acknowledge the support of NSF and AFOSR during the period in which this manuscript was written.

Notation

$\{\}$	The empty set.
(X, τ)	The set X endowed with the topology τ.
(X, d)	The set X endowed with the metric d.
$(X, \|\cdot\|)$	The set X endowed with the norm $\|\cdot\|$.
R	The real number system.
R^n	The n dimensional Euclidean space.
$|x|_p$	The p-norm of the vector $x \in R^n$ defined as $|x|_p := \left(\sum_{i=1}^{n} |x_i|^p\right)^{\frac{1}{p}}$.
$|x|_1$	The 1-norm of the the vector $x \in R^n$.
$|x|_2$	The 2-norm of the the vector $x \in R^n$.
$\hat{x}(\lambda)$	The λ transform of a right sided real sequence $x = (x(k))_{k=0}^{\infty}$ defined as $\hat{x}(\lambda) := \sum_{k=0}^{\infty} x(k)\lambda^k$.
ℓ	The vector space of sequences.
$\ell^{m \times n}$	The vector space of matrix sequences of size $m \times n$..
ℓ_1	The Banach space of right sided absolutely summable real sequences with the norm given by $\| x \|_1 := \sum_{k=0}^{\infty} |x(k)|$.
$\ell_1^{m \times n}$	The Banach space of matrix valued right sided real sequences with the norm $\|x\|_1 := \max_{1 \leq i \leq m} \sum_{j=1}^{n} \|x_{ij}\|_1$ where $x \in \ell_1^{m \times n}$ is the matrix (x_{ij}) and each x_{ij} is in ℓ_1.
ℓ_∞	The Banach space of right sided, bounded sequences with the norm given by $\| x \|_\infty := \sup_k |x(k)|$.
$\ell_\infty^{m \times n}$	The Banach space of matrix valued right sided real sequences with the norm $\|x\|_\infty := \sum_{i=1}^{m} \max_{1 \leq j \leq n} \|x_{ij}\|_\infty$ where $x \in \ell_\infty^{m \times n}$ is the matrix (x_{ij}) and each x_{ij} is in ℓ_∞.
c_0	The subspace of ℓ_∞ with elements x that satisfy $\lim_{k \to \infty} x(k) = 0$.
$c_0^{m \times n}$	The Banach subspace of $\ell_\infty^{m \times n}$ with elements x which satisfy $\lim_{k \to \infty} x(k) = 0$.
ℓ_2	The Banach space of right sided square summable sequences with the norm given by $\| x \|_2 := \left(\sum_{k=0}^{\infty} x(k)^2\right)^{\frac{1}{2}}$.
$\ell_2^{m \times n}$	The Banach space of matrix valued right sided real sequences with the norm $\|x\|_2 := \left(\sum_{i=1}^{m} \sum_{j=1}^{n} \|x_{ij}\|_2^2\right)^{\frac{1}{2}}$ where $x \in \ell_2^{m \times n}$ is the matrix (x_{ij}) and each x_{ij} is in ℓ_2.

$\mathcal{H}_2$	The isometric isomorphic image of ℓ_2 under the λ transform $\hat{x}(\lambda)$ with the norm given by $\| \hat{x}(\lambda) \|_2 = \| x \|_2$.
X^*	The dual space of the Banach space X. $< x, x^* >$ denotes the value of the bounded linear functional x^* at $x \in X$.
$W(X^*, X)$	The weak star topology on X^* induced by X.
T^*	The adjoint operator of $T : X \to Y$ which maps Y^* to X^*.
$int(A)$	The interior of the set A.
$\mathcal{D}$	The closed unit disc in the complex plane.
A'	The transpose of the matrix A.

Contents

1. Topology

In this chapter we lay down the foundations of the mathematical structure required for optimization methods for vector spaces. We start by a terse introduction to sets. No attempt is made to provide an axiomatic description of set theory. We appeal to the intuitive notion of sets as being a collection of objects. The reader is introduced to the axiom of choice and Zorn's lemma. The meat of this chapter is the section on general topology where we study topological sets with the bare minimum of structure on the sets. The concepts of convergence, continuity and compactness are presented in this general setting.

The next two sections discuss metric and vector normed spaces. Vector normed spaces will be studied in greater detail in the next chapter. The section on finite dimensional spaces summarizes some important properties enjoyed by finite dimensional spaces that are lacking in infinite dimensional spaces.

In the last section of this chapter we study extrema of real valued functions. It is shown that compact sets and various forms of continuity play a pivotal role in the existence of extrema. Thus they form the focus of optimization. It is shown that the norm topology does not have an abundance of compact sets.

Elementary knowledge of real analysis is assumed. Definitions for well known operations like union and intersection of sets is not provided. Familiarity with countable sets, uncountable sets and the real number system is assumed. Except for these topics this chapter is self contained. However, the material presented in this chapter will be more transparent to a reader who has been exposed to analysis concepts taught in a undergraduate course (the first seven chapters of [1] is sufficient background).

1.1 Sets

Here, we do not attempt to provide the axiomatic development of sets, rather we appeal to the intuitive notion of a set as being a collection of objects. If A and B are two sets then $A \times B$ is another set defined as $A \times B := \{(a,b) : a \in A \text{ and } b \in B\}$. A *binary relation* on a set X is a subset T of $X \times X$.

The relation is usually denoted by a symbol $\prec$ and we say $x \prec y$ if and only if (x, y) belongs to T.

An *order* is a binary relation which is transitive ($x \prec y$ and $y \prec z$ implies that $x \prec z$), reflexive ($x \prec x$) and antisymmetric ($x \prec y$ and $y \prec x$ implies that x is the same element as y). A relation without the antisymmetric property is called a *preorder.*

An element x is called the *majorant* of the subset Y if for all y in Y, $y \prec x$. An element, x is called the *minorant* of the subset Y if for all y in Y, $x \prec y$. The set is said to be *totally ordered* if either $x \prec y$ or $y \prec x$ for any two elements x and y of X. Furthermore, it is *well ordered* if every subset Y of X has a minorant in the set Y.

A *directed set* X is a set with a preorder such that every pair (x, y) of X has a majorant. We say that $(X, \prec)$ is *inductively ordered* if each totally ordered subset of X (in the order induced from X) has a majorant in X. We denote the collection of all subsets of a set X by $\chi(X)$ and the empty set by $\{\}$. For sets A and B, we define $A \backslash B$ as the collection of all elements of A which are not in B. The *axiom of choice* states that it is possible to choose an element from any nonempty subset of a set. We now state this axiom precisely.

Axiom 1.1 (Axiom of choice) *Given any set X there exists a function c such that*

$$c : (\chi(X) \setminus \{\}) \to X.$$

c is called the choice function.

In the intuitive description of a set, thought of as a collection of elements, it is difficult to explain the role of the axiom of choice. In the axiomatic description of set theory it can be shown that the axiom of choice is independent of the axioms used in developing set theory. Thus, different mathematics results based on whether the axiom of choice is accepted to be true or not. The axiom of choice results in important theorems (such as the Hahn-Banach theorem). There seems to be no other alternative in establishing such key results and thus the validity of the axiom of choice is generally accepted.

Lemma 1.1.1 (Zorn's lemma). *Every inductively ordered set X has an element x such that if y is in X and $x \prec y$ then $y = x$. x is called the maximal element.*

It can be shown that Zorn's lemma is equivalent to the axiom of choice. This lemma will be used in obtaining important results.

1.2 General Topology

Definition 1.2.1 (Topology). *A topology on a set X is a collection of subsets τ of X with the following properties.*

1. *Any union of sets in τ belongs to τ.*
2. *Any finite intersection of sets in τ belongs to τ.*
3. *$\{\}$ and X belong to τ.*

We say that (X, τ) is a *topological set* and that τ consists of the *open* subsets of X. A subset F of a topological set (X, τ) is said to be *closed* if $X \backslash F$ is open. A subset Y in a topological set (X, τ) is a *neighbourhood* of the point x in X if there is an open subset A of Y such that $x \in A$. For each subset Y of X we define the *closure* of Y as the intersection of all the closed sets that contain Y. The closure of the set Y is denoted by Y^-.

Theorem 1.2.1. *For a topological set (X, τ), the following assertions are true.*

1. *The union of any collection of open sets is open and the union of a finite collection of closed sets is closed.*
2. *The intersection of any collection of closed sets is closed and the intersection of a finite collection of open sets is open.*
3. *If $Y \subset X$ then $x \in Y^-$ if and only if for every neighbourhood A of x, $Y \cap A \neq \{\}$.*

Proof. (3.)

$$\begin{aligned} A^- &= \bigcap \{F : F \supset A \text{ and } F \text{ is a closed set}\} \\ &= \bigcap \{X \backslash G : X \backslash G \supset A \text{ and } G \in \tau\}. \end{aligned}$$

Therefore x_0 belongs to A^- if and only if $x_0 \in X \backslash G$ for every open set G which is such that $X \backslash G \supset A$. Therefore x_0 is in A^- if and only if $x_0 \in X \backslash G$ for every open set G which is such that $G \cap A = \{\}$. This is equivalent to the statement that x_0 is in A^- if and only if for all G open which contain x_0, $G \cap A \neq \{\}$. We leave the rest of the proof to the reader. □

Definition 1.2.2 (Relative topology). *If (X, τ) is a topological set and Y is a subset of X then the relative topology on Y is a collection of sets of the form $Y \cap A$ where A belongs to τ.*

It follows from the definition of relative topology that a set is closed in the relative topology on Y if and only if it has the form $Y \cap F$ where F is closed in (X, τ).

Definition 1.2.3 (Interior of a set). *Let Y be a subset of a topological set (X, τ). A point x is called an interior point of Y if there is an open set $A \in \tau$ such that $x \in A$ with $A \subset Y$. The collection of all interior points of the subset Y is called the interior of the set Y and is denoted by int(Y).*

Given a topological set (X, τ) and a point $x \in X$ we define the *neighbourhood filter*, $\mathcal{N}(x)$ associated with x by

$$\{N : N \subset X \text{ such that there exists } U \in \tau, x \in U \text{ and } U \subset N\}, \tag{1.1}$$

which is a collection of neighbourhoods of x. We say that for any two sets A and B in $\mathcal{N}$, $A \prec B$ if $A \supset B$. It is can be shown that the relation $\prec$ defined above is reflexive and transitive. Also, given A and B in $\mathcal{N}(x)$, $A \cap B$ is in $\mathcal{N}(x)$ and is contained in A and B. Therefore, any two elements in $\mathcal{N}(x)$ have a majorant. This implies that $\mathcal{N}(x)$ with the relation $\prec$ is a directed set.

If σ and τ are two topologies on X then we say that σ is a *stronger topology* than τ or τ is a *weaker topology* than σ if $\tau \subset \sigma$.

Lemma 1.2.1. *Let $\{\tau_j : j \in \Lambda\}$ be a collection of topologies on a set X indexed by the set Λ. Then there exists a weakest topology which contains all τ_j for j in Λ. There also exists a strongest topology which is contained in all τ_j for j in Λ.*

Proof. Let τ_{int} be the collection of sets which are in all τ_j for j in Λ. Then it is clear that τ_{int} defines a topology on X. Also, τ_{int} is contained in all τ_j for j in Λ. It is also easy to see that τ_{int} is the strongest such topology.

Let $\mathcal{T}$ be the collection of all topologies which are stronger than τ_j for all j in Λ. This collection is not empty because the topology in which all elements of X are defined open (called the discrete topology) is in $\mathcal{T}$. We know that there exists a strongest topology τ_{out} which is contained in all topologies in $\mathcal{T}$. This is the weakest topology which contains all τ_j for j in Λ. □

Definition 1.2.4 (Subbase, Base, Neighbourhood base). *Let σ be a collection of subsets of X. Then σ is a subbase for the topology τ, if τ is the weakest topology, amongst topologies on X that contain σ. σ is a basis for τ if all sets in it are unions of elements in σ. σ is a neighbourhood basis for an element x in (X, τ) if for every set A in the neighbourhood filter of x (given by $\mathcal{N}(x)$) there exists a set in both, $\mathcal{N}(x)$ and σ which is a subset of A.*

Lemma 1.2.2. *The following assertions are true.*

1. *Let σ be a collection of subsets of a set X. Let τ be a topology on X. σ is a subbasis for τ if and only if τ is the set $\{F : F = X,\ F = \{\}$ or F is a union of sets which are finite intersections of sets in $\sigma\}$ for which σ is a subbasis. Then every set in $\tau(\sigma)$ is a union of sets which are finite intersections of sets in σ, or the null set or the set X.*
2. *A collection of open sets σ in a topological set (X, τ) is a basis for τ if σ forms a neighbourhood basis for all elements in X.*
3. *A collection of open sets σ in a topological set (X, τ) is a subbasis for τ if the collection of sets formed by finite intersections of sets in σ forms a neighbourhood basis for all elements in X.*

Proof. (1.) Every topology that contains σ contains all the sets which are union of sets formed by finite intersections of sets in σ. This follows from the definition of a topology on a set. Also, all sets which are unions of sets formed by finite intersection of sets from σ together with the null set and the set X define a topology on X.

(2.) Let $A \in \tau$. Then for every x in A, A is in $\mathcal{N}(x)$ where $\mathcal{N}(x)$ is the neighbourhood filter of x. As the collection of sets σ forms a neighbourhood basis for all points in X, it follows that there exists a set B_x such that $B_x \in \sigma \cap \mathcal{N}(x)$ and $B_x \subset A$. It is evident that $A = \bigcup_{x \in A} B_x$. Therefore, A can be written as a union of sets from σ. As A is an arbitrary set in τ it follows that that σ forms a basis for the topology τ.

(3.) Let $A \in \tau$ be such that $A \neq X$ and $A \neq \{\}$. Then for every x in A , A is in $\mathcal{N}(x)$ where $\mathcal{N}(x)$ is the neighbourhood filter of x. It follows that there exists a set B_x which is a finite intersection of sets from σ such that $B_x \in \mathcal{N}(x)$ and $B_x \subset A$. It is evident that $A = \bigcup_{x \in A} B_x$. Therefore, A can be written as a union of sets from the collection of sets formed by finite intersections of sets from σ. As A is an arbitrary set in τ it follows that σ forms a subbasis for the topology τ.

□

Definition 1.2.5 (Nets). *A net in a set X is a pair (Λ, k) where Λ is a directed set and k is a map from Λ into X. A net is also denoted as $\{x_\lambda\}_{\lambda \in \Lambda}$ where $x_\lambda = k(\lambda)$. We also say that x_λ is a net in X on Λ.*

The net x_λ is *eventually* in the set A if there exists $\lambda_0 \in \Lambda$ such that if $\lambda_0 \prec \lambda$ then $x_\lambda \in A$. The net x_λ is *frequently* in the set A if for every $\lambda_0 \in \Lambda$ there exists a $\lambda \in \Lambda$ such that $\lambda_0 \prec \lambda$ and $x_\lambda \in A$.

Definition 1.2.6 (Convergence). *Let x_λ be a net in a topological set (X, τ). We say that $x_\lambda \to x_0$ if x_λ is eventually in N for all $N \in \mathcal{N}(x_0)$. We also say that x_λ converges to x_0 in the topology τ. Another notation used is* $\lim x_\lambda$ *to represent x_0 and x_0 is said to be the limit of the net x_λ.x_λ is said to be a convergent net if there exists a $x_0 \in X$ such that $x_\lambda \to x_0$.*

It is possible that a net converges to more than a single element. This is not the case for the *Hausdorff* topology which is defined below.

Definition 1.2.7 (Hausdorff topology). *For a set X, τ is a Hausdorff topology if for all elements x and y in X with $x \neq y$ there exist sets $A \in \tau$ and $B \in \tau$ such that $x \in A$, $y \in B$ and $A \cap B$ is empty.*

Definition 1.2.8 (Denseness, Separability, Axiom of countability). *A subset Y of a topological set (X, τ) is dense in (X, τ) if $Y^- = X$. The topological set (X, τ) is called separable if it has a countable dense subset. The topological set (X, τ) is said to satisfy the first axiom of countability if for every x in X there exists a countable number of open sets $A_n(x)$ such that any neighbourhood of x contains atleast one of them.*

A net defined on the set of integers in X is called a *sequence*. In most results the concept of a net is superfluous and can be replaced by the notion of a sequence if the topological set satisfies the first axiom of countability. However, the concept of nets allows for more general results and the proofs of many results become easier to establish.

Lemma 1.2.3. *Let (X, τ) be a topological set and let A be a subset of X. Then x_0 is in the closure of A if and only if there is a net x_λ in A such that $x_\lambda \to x_0$.*

Proof. From Theorem 1.2.1 we know that if $A \subset X$, then $x_0 \in A^-$ if and only if for every neighbourhood G of x_0, $A \cap G \neq \{\}$.

($\Leftarrow$) Suppose there is a net x_λ in A such that $x_\lambda \to x_0$. Then for any open set G which contains x_0 (and therefore is in $\mathcal{N}(x_0)$), there exists a λ_0 such that $\lambda_0 \prec \lambda$ implies that $x_\lambda \in G$. However, every such x_λ is in A because x_λ is a net in A. Therefore $G \cap A \neq \{\}$ which implies that $x_0 \in A^-$.

($\Rightarrow$) Suppose $x_0 \in A^-$ and $N \in \mathcal{N}(x_0)$. Then there exists $G \in \tau$ such that $G \subset N$ and $x_0 \in G$. Therefore for every $N \in \mathcal{N}(x_0), N \cap A \neq \{\}$. Let x_N belong to $N \cap A$. x_N is a net in A defined on the directed set $\mathcal{N}(x_0)$ (we have shown before that $\mathcal{N}(x_0)$ is a directed set with $A \prec B$ if and only if $A \supset B$). Given any $N \in \mathcal{N}(x_0)$, let $\lambda_0 = N$. If $\lambda_0 \prec \lambda$ then $x_\lambda \in N$ because $N = \lambda_0 \supset \lambda$ and $x_\lambda \in \lambda$. Thus x_N is a net in A such that $x_N \to x_0$. This proves the lemma. □

Example 1.2.1. It is not true that if x_0 is in the closure of a set A then there exists a sequence in A which converges to x_0. Consider any uncountable set X and define the topology on X by

$$\tau := \{\{\}, X, \text{all sets which have countable complements}\}.$$

Then the closed sets are given by $\{\{\}, X, \text{all countable sets}\}$. Let $A := X \backslash \{x_0\}$ where x_0 is any element in X. Then A is open and X is the only closed set that contains A. Therefore $A^- = X$. Let x_n be any sequence in A. The set $X \backslash \{x_n : n \geq 1\}$ is an open set which contains x_0 and therefore is a neighbourhood of x_0. However x_n is never in this neighbourhood and therefore $x_n \not\to x_0$.

This example illustrates that to fully describe topological concepts in terms of convergence, the use of nets is indispensable.

Definition 1.2.9 (Continuity). *Let (X, τ) and (Y, σ) be topological sets. A function $f : (X, \tau) \to (Y, \sigma)$ is continuous if for every set $G \in \sigma$, $f^{-1}(G) \in \tau$. $f^{-1} : (Y, \sigma) \to (X, \tau)$ is defined as*

$$f^{-1}(B) = \{x \in X : f(x) \in B\}.$$

f is said to be continuous at a point x_0 if for every neighbourhood G of $f(x_0)$, $f^{-1}(G)$ is a neighbourhood of x_0.

Lemma 1.2.4. *The following statements are equivalent.*

1. $f : (X, \tau) \to (Y, \sigma)$ is continuous at every point $x \in X$.

2. $f:(X,\tau)\to(Y,\sigma)$ *is a continuous function.*
3. If x_λ *is a net such that* $x_\lambda\to x_0$ *in* (X,τ) *then* $f(x_\lambda)\to f(x_0)$ *in* (Y,σ).

Proof. ($1\Rightarrow 2$) Suppose $f:(X,\tau)\to(Y,\sigma)$ is continuous at every point $x\in X$. Let $G\in\sigma$ and let $x\in f^{-1}(G)$. As G is a neighbourhood of $f(x)$ we have that $f^{-1}(G)$ is a neighbourhood of x. This implies that for every $x\in f^{-1}(G)$ there exists a set A_x in τ containing x such that $A_x\subset f^{-1}(G)$. It is easy to show that

$$\bigcup\{A_x : x\in f^{-1}(G)\}=f^{-1}(G).$$

Therefore $f^{-1}(G)$ (being a union of open sets) is open. As G is an arbitrary open set in Y we have shown that f is continuous.

($2\Rightarrow 1$) Suppose f is a continuous function. Let N be a neighbourhood of $f(x)$ for some $x\in X$. Then there exists a set $G\in\sigma$ such that $f(x)\in G$ and $G\subset N$. From continuity of f we have that $f^{-1}(G)\in\tau$. As $f^{-1}(N)\supset f^{-1}(G)$ and $f^{-1}(G)$ is open and contains x we have that $f^{-1}(N)$ is a neighbourhood of x. Therefore we have shown that for any neighbourhood N of $f(x)$, $f^{-1}(N)$ is a neighbourhood of x. As x was chosen arbitrarily we have that f is continuous at every point $x\in X$.

($2\Rightarrow 3$) Suppose f is continuous. Given any neighbourhood N of $f(x_0)$ we know that there exists a set $G\in\sigma$ such that $f(x_0)\in G$ and $G\subset N$. From continuity of f we know that $f^{-1}(G)\in\tau$. Also, $x_0\in f^{-1}(G)$. Therefore $f^{-1}(G)$ is a neighbourhood of x_0. As $x_\lambda\to x_0$ we know that there exists a λ_0 such that $\lambda_0\prec\lambda$ implies that x_λ is in $f^{-1}(G)$. Therefore there exists a λ_0 such that $\lambda_0\prec\lambda$ implies that $f(x_\lambda)\in G\subset N$. As the neighbourhood N of $f(x_0)$ was chosen arbitrarily we have shown that $f(x_\lambda)\to f(x_0)$.

($3\Rightarrow 2$) Suppose f is not continuous. Then there exists $x_0\in X$ such that f is not continuous at x_0 (from $2\Rightarrow 1$). This implies that there exists a neighbourhood N of $f(x_0)$ such that $f^{-1}(N)$ is not a neighbourhood of x_0. Therefore given $M\in\mathcal{N}(x_0)$ there exists $x_M\in M$ such that $x_M\notin f^{-1}(N)$. x_M is a net in X on the directed set $\mathcal{N}(x_0)$ where the binary relation is given by $A\prec B$ if and only if $A\supset B$. It is clear that $x_M\to x_0$. However, $f(x_M)\not\to f(x_0)$ because $f(x_M)\in Y\backslash N$ for all $M\in\mathcal{N}(x_0)$ and N is a neighbourhood of $f(x_0)$. This proves the lemma. □

For a map f between sets X and Y and $A\subset X$ we define

$$f(A):=\{y\in Y : y=f(x) \text{ for some } x\in A\}.$$

The set $f(X)$ is called the range of f and is denoted by *range*(f).

Lemma 1.2.5. *If* $f:(X,\tau_x)\to(Y,\tau_y)$ *is a continuous map then* $f^{-1}(B)$ *is closed if* B *is closed in* Y.

Proof. As B is closed in Y it follows that $Y\backslash B$ is open. From continuity of f it follows that $f^{-1}(Y\backslash B)$ is open. However, $f^{-1}(Y\backslash B)=X\backslash f^{-1}(B)$. Therefore, $f^{-1}(B)$ is closed. □

Definition 1.2.10 (Initial topology). *Let $\{f_j : j \in J\}$ be a collection of functions, indexed by the set J on a set X such that $f_j : X \to Y_j$ where Y_j is a topological set with topology τ_j. The weakest topology on X which makes all functions f_j continuous is called the initial topology on X induced by f_j.*

Lemma 1.2.6. *Let $\{f_j : j \in J\}$ be a collection of functions on a set X, indexed by the set J, such that $f_j : X \to Y_j$ where Y_j is a topological set with topology τ_j. The collection of sets of the form*

$$\{f_j^{-1}(A) : A \in \tau_j \textit{ for some } j \in J\},$$

forms a subbasis for the initial topology on X.

Proof. It is clear that any topology on X which makes all the functions f_j continuous must contain all the sets from σ where

$$\sigma := \{f_j^{-1}(A) : A \in \tau_j \text{ for some } j \in J\},$$

otherwise at least one of the f_j will be discontinuous. Therefore, the initial topology also contains σ. Let τ denote the topology for which σ forms the subbasis. As τ is the weakest toplogy that contains σ and the initial topology contains σ it follows that the initial topology is stronger than τ. Also if X is endowed with the topology τ then the functions f_j are continuous. As the initial topology is the weakest topology on X which make f_j continuous for all j it follows that τ is stronger than the initial topology. Thus τ is the initial topology. □

Lemma 1.2.7. *Let X be endowed with the initial topology τ induced by a collection of functions $\{f_j : j \in J\}$, indexed by the set J, where $f_j : X \to (Y_j, \tau_j)$. Then a net x_λ in (X, τ) on Λ converges to x_0 in X if and only if $f_j(x_\lambda)$ converges to $f_j(x_0)$ in τ_j for all j in J.*

Proof. ($\Rightarrow$) Suppose the net x_λ in X on Λ converges to x_0. Then, because the topology on X is the initial topology (which makes all f_j continuous) $f_j(x_\lambda) \to f_j(x_0)$ for all $j \in J$.

($\Leftarrow$) Suppose for all $j \in J$, $f_j(x_\lambda) \to f_j(x_0)$. Let B be a neighbourhood of x_0. Then, as

$$\sigma := \{f_j^{-1}(A) : A \in \tau_j \text{ for some } j \in J\},$$

is a subbasis for the initial topology on X, B contains a set C which is a finite intersection of sets from σ such that $x_0 \in C$. Without loss of generality assume that

$$C = \cap_{i=1}^{k}\{f_i^{-1}(A_i) : \text{ where } A_i \in \tau_i\}.$$

Note that as $x_0 \in C$ it follows that A_i is an open set containing $f_i(x_0)$ for all $i = 1, \ldots, k$. As $f_i(x_\lambda) \to f_i(x_0)$ it follows that there exist λ_i such that if $\lambda_i \prec \lambda$ then $f_i(x_\lambda) \in A_i$ for all $i = 1, \ldots, k$. Let λ_0 represent the majorant of the set $\{\lambda_1, \ldots, \lambda_k\}$. This implies that if $\lambda_0 \prec \lambda$ then $f_i(x_\lambda) \in A_i$ for all

$i = 1, \ldots, k$. Therefore, if $\lambda_0 \prec \lambda$ then $x_\lambda \in f_i^{-1}(A_i)$ for all $i = 1, \ldots, k$ which implies that $x_\lambda \in \cap_{i=1}^{k}\{f_i^{-1}(A_i)\} \subset B$. This implies that x_λ is eventually in B. As B is an arbitrary neighbourhood of x_0 it follows that $x_\lambda \to x_0$. □

Definition 1.2.11 (Isomorphism). *A function $f : X \to Y$ is an isomorphism if it is a one-to-one and onto function.*

Definition 1.2.12 (Homeomorphism). *A function $f : (X, \tau_x) \to (Y, \tau_y)$ is a homeomorphism if it is a one-to-one and an onto function which is continuous with its inverse also continuous.*

Two topological sets which have a homeomorphism between them are said to be topologically identical.

Definition 1.2.13 (Filters). *A filter $\mathcal{F}$ in a set X is collection of subsets of X which has the following properties.*

1. $\{\} \notin \mathcal{F}$.
2. $X \in \mathcal{F}$.
3. *$A \subset B$ and $A \in \mathcal{F}$ implies that $B \in \mathcal{F}$.*
4. *$A \in \mathcal{F}$ and $B \in \mathcal{F}$ implies that $A \cap B \in \mathcal{F}$.*

Definition 1.2.14 (Ultrafilter). *An ultrafilter in a set X is a filter in X with the additional property that no other filter in X properly contains it.*

Lemma 1.2.8. *For a filter $\mathcal{G}$ in a set X, the following statements are equivalent.*

1. *$\mathcal{G}$ is an ultrafilter in X.*
2. *For every set $A \subset X$ either $A \in \mathcal{G}$ or $X \backslash A \in \mathcal{G}$.*

Proof. ($2 \Rightarrow 1$) Suppose $\mathcal{G}$ is not an ultrafilter in X. Then there exists a filter $\mathcal{F}$ in X and a subset A of X such that $A \in \mathcal{F}$ but $A \notin \mathcal{G}$ with the property that if $B \in \mathcal{G}$ then $B \in \mathcal{F}$. If $X \backslash A \in \mathcal{G}$ then $X \backslash A \in \mathcal{F}$ and because $\mathcal{F}$ is a filter, $(X \backslash A) \cap A \in \mathcal{F}$. This would imply that $\{\} \in \mathcal{F}$. Therefore $X \backslash A \notin \mathcal{G}$. Therefore we have shown that both A and $X \backslash A$ are not in the filter $\mathcal{G}$.

($1 \Rightarrow 2$) Suppose $\mathcal{G}$ is a filter and there exists a set $Y \subset X$ such that $Y \notin \mathcal{G}$ and $X \backslash Y \notin \mathcal{G}$. Let A be any set which belongs to the filter $\mathcal{G}$. Then $(X \backslash Y) \cap A \neq \{\}$ because otherwise $A \subset Y$ and as $\mathcal{G}$ is a filter $Y \in \mathcal{G}$. Let

$$\mathcal{F} := \{C : \text{ there exists } A \in \mathcal{G} \text{ such that } C \supset (X \backslash Y) \cap A\}.$$

As $(X \backslash Y) \cap A \neq \{\}$ for every $A \in \mathcal{G}$ it is clear that $\{\} \notin \mathcal{F}$. It is clear that $X \in \mathcal{F}$. If $C_1 \in \mathcal{F}$ then there exists a set $A_1 \in \mathcal{G}$ such that $C_1 \supset (X \backslash Y) \cap A_1$. If $C_2 \supset C_1$ then $C_2 \supset (X \backslash Y) \cap A_1$ and therefore $C_2 \in \mathcal{F}$.

If C_1 and C_2 both belong to $\mathcal{F}$ then there exist sets A_1 and A_2 both in $\mathcal{G}$ such that $C_1 \supset (X \backslash Y) \cap A_1$ and $C_2 \supset (X \backslash Y) \cap A_2$. As $A_1 \cap A_2 \in \mathcal{G}$ and $C_1 \cap C_2 \supset (X \backslash Y) \cap (A_1 \cap A_2)$ it follows that $C_1 \cap C_2 \in \mathcal{F}$. This proves that $\mathcal{F}$ is a filter. Note that if $B \in \mathcal{G}$ then $B \in \mathcal{F}$. Also, $X \backslash Y \in \mathcal{F}$ whereas $X \backslash Y \notin \mathcal{G}$. Thus $\mathcal{F}$ properly contains $\mathcal{G}$. Therefore $\mathcal{G}$ is not an ultrafilter, which proves the lemma. □

Lemma 1.2.9. *Every filter in a set X is contained in an ultrafilter in X.*

Proof. Suppose $\mathcal{G}$ is a filter in X. Consider

$$P := \{\mathcal{H} : \mathcal{H} \text{ is a filter in } X \text{ which contains } \mathcal{G}\}.$$

Let the binary relation on P be given by $\prec$ where $\mathcal{B} \prec \mathcal{A}$ if and only if $\mathcal{B} \subset \mathcal{A}$. Let Q be a totally ordered subset of P. Let N denote the union of all the elements of Q. Then it can be shown that $N \in P$ and that N is a majorant for Q. This implies that P is inductively ordered. From Zorn's lemma (Lemma 1.1.1) P has a maximal element $\mathcal{F}$. It follows that $\mathcal{F}$ is an ultrafilter that contains $\mathcal{G}$. □

Definition 1.2.15 (Subnets). *Let (Λ, i) be a net in X on Λ. A subnet of (Λ, i) is a net (M, j) in X with a function $h : M \to \Lambda$ such that $j = i(h)$ and for every $\lambda_0 \in \Lambda$ there exists a $\beta_0 \in M$ with $\lambda_0 \prec h(\beta)$ if $\beta_0 \prec \beta$. We also say that y_β on M is a subnet of the net x_λ on Λ if there exists a function $h : M \to \Lambda$ such that $x_{h(\beta)} = y_\beta$ and for every $\lambda_0 \in \Lambda$ there exists a $\beta_0 \in M$ with $\lambda_0 \prec h(\beta)$ if $\beta_0 \prec \beta$.*

The definition of subnets seems involved. However, the definition of a subsequence of a sequence will bring out the similarity between subnets and subsequences.

Lemma 1.2.10. *Let x_λ be a net in (X, τ) on Λ. Let y_β be a subnet of x_λ on M. If $x_\lambda \to x_0$ then $y_\beta \to x_0$.*

Proof. Let $N \in \mathcal{N}(x_0)$. As $x_\lambda \to x_0$ we know that there exists a $\lambda_0 \in \Lambda$ such that $\lambda_0 \prec \lambda$ implies that $x_\lambda \in N$. As y_β is a subnet of x_λ there exists a function $h : M \to \Lambda$ and $\beta_0 \in M$ such that $\beta_0 \prec \beta$ implies that $\lambda_0 \prec h(\beta)$. Thus there exists a $\beta_0 \in M$ such that if $\beta_0 \prec \beta$ then $x_{h(\beta)} \in N$. But from the definition of a subnet $x_{h(\beta)} = y_\beta$ and therefore we have shown that for every $N \in \mathcal{N}(x_0)$ there exists a $\beta_0 \in M$ such that if $\beta_0 \prec \beta$ then $y_\beta \in N$. □

Definition 1.2.16 (Universal nets). *A net x_λ in X on Λ is universal if for all $A \subset X$ either x_λ is eventually in A or x_λ is eventually in $X \backslash A$.*

As we will see universal nets play an important role in characterizing compactness. The following lemma brings out an interesting property of universal nets.

Lemma 1.2.11. *If x_λ is a universal net in X on Λ and $f : X \to Y$ where Y is a set then $f(x_\lambda)$ is a universal net in Y on Λ.*

Proof. Let $B \subset Y$. Then as x_λ is a universal net on X it follows that x_λ is eventually either in $f^{-1}(B)$ or in $X \backslash f^{-1}(B) = f^{-1}(Y \backslash B)$. Therefore, $f(x_\lambda)$ is eventually either in B or in $Y \backslash B$. Therefore, $f(x_\lambda)$ is a universal net in Y. □

Theorem 1.2.2. *Every net has a universal subnet.*

Proof. Let x_λ be a net in X on Λ. Consider the set

$$\mathcal{G} := \{A : A \subset X \text{ and } x_\lambda \text{ is eventually in } A\}.$$

It is evident that $\mathcal{G}$ is a filter. From Lemma 1.2.9 we know that there exists an ultrafilter $\mathcal{F}$ which contains $\mathcal{G}$.

We will show now that x_λ is frequently in every set in $\mathcal{F}$. Suppose there exists a set $F \in \mathcal{F}$ and a $\lambda_0 \in \Lambda$ such that if $\lambda_0 \prec \lambda$ then $x_\lambda \notin F$. Therefore $\{x_\lambda : \lambda_0 \prec \lambda\} \cap F = \{\}$. The set $\{x_\lambda : \lambda_0 \prec \lambda\}$ belongs to $\mathcal{G}$ and therefore it belongs to $\mathcal{F}$. F belongs to $\mathcal{F}$ and as $\mathcal{F}$ is a filter it follows that $\{x_\lambda : \lambda_0 \prec \lambda\} \cap F \in \mathcal{F}$. This means that the null set belongs to $\mathcal{F}$. We have reached a contradiction and therefore x_λ is frequently in every set in $\mathcal{F}$.

Now we construct an universal subnet of x_λ. Consider the set

$$M := \{(\lambda, B) : \lambda \in \Lambda \text{ and } B \in \mathcal{F}\}$$

with the order defined by $(\lambda_1, B_1) \prec (\lambda_2, B_2)$ if and only if $\lambda_1 \prec \lambda_2$ in Λ and $B_1 \supset B_2$. It follows that M with the relation $\prec$ defined above is a directed set. Define the mapping $h : M \to \Lambda$ so that $\lambda \prec h(\lambda, B)$ and $x_{h(\lambda,B)} \in B$ (such a map exists because x_λ is frequently in every set of $\mathcal{F}$). Let the elements of M be denoted by β and define $y_\beta = x_{h(\lambda,B)}$ where $\beta = (\lambda, C)$. Given any $\lambda_0 \in \Lambda$ choose $\beta_0 \in M$ such that $\beta_0 = (\lambda_0, X)$. Then $\beta_0 \prec \beta = (\lambda, C)$ implies that $\lambda_0 \prec \lambda$ in Λ. Therefore given $\lambda_0 \in \Lambda$ there exists a $\beta_0 \in M$ such that $\beta_0 \prec \beta$ implies that $\lambda_0 \prec \lambda \prec h(\lambda, C)$ and therefore y_β is a subnet of x_λ.

Given any subset A of X we know that either $A \in \mathcal{F}$ or $X \backslash A \in \mathcal{F}$ (see Lemma 1.2.8). Suppose that $A \in \mathcal{F}$. Let $\beta_0 = (\lambda_0, A)$ where $\lambda_0 \in \Lambda$ is arbitrary. If $\beta = (\lambda, C)$ is such that $\beta_0 \prec \beta$ then $\lambda_0 \prec \lambda$ and $A \supset C$. Also, $y_\beta = x_{h(\lambda,C)} \in C$ and therefore $y_\beta \in A$. Therefore we have shown that there exists a $\beta_0 \in M$ such that $\beta_0 \prec \beta$ implies that $y_\beta \in A$. Thus the net y_β is eventually in the set A. It can be shown in a similar manner that if $X \backslash A \in \mathcal{F}$ then y_β is eventually in $X \backslash A$. As A was chosen arbitrarily we have shown that y_β is a universal subnet of x_λ. □

Definition 1.2.17 (Open cover). *An open covering of a subset Y of a topological set (X, τ) is a subset σ of τ such that $Y \subset \bigcup A, A \in \sigma$. A subcovering of σ is a covering σ_0 that is contained in σ.*

Definition 1.2.18 (Compactness). *A topological set (X, τ) is compact if every open cover of X has a finite subcover.*

Thus (X, τ) is compact if for every collection (countable or otherwise) of open sets which cover X there is a finite number from the collection which cover X. A subset of (X, τ) is said be compact if it is compact in the relative topology.

Lemma 1.2.12. *Every closed subset C of a compact topological set (X, τ) is compact.*

Proof. Suppose, $\{A_\alpha\}$ forms an open cover for C. Then, there exists G_α in τ such that $G_\alpha \cap C = A_\alpha$. As C is closed we have that $G_0 := X \backslash C$ is open. Therefore $\{G_\alpha\}$ together with G_0 forms an open cover of X. As X is compact there exists a finite number of sets from G_0 and $\{G_\alpha\}$ which cover X. It follows that only a finite number of sets from $\{G_\alpha\}$ and G_0 are required to cover X. This implies that only finitely many of the the sets $\{G_\alpha\}$ are required to cover C. Thus we have shown that given any open cover $\{A_\alpha\}$ only finitely many of the sets from $\{A_\alpha\}$ form an open cover of C. This proves the lemma. □

Not every compact subset of a closed set is closed. However, the following lemma holds.

Lemma 1.2.13. *Every compact subset of a closed set in a topological set is closed if the topology is Hausdorff.*

Lemma 1.2.14. *If $f : (X, \tau_x) \to (Y, \tau_y)$ is a continuous map from a compact topological set (X, τ_x) to a topological set (Y, τ_y) then $f(X)$ is compact as a subset of (Y, τ_y) where*

$$f(X) := \{y : \text{ there exists a } x \in X \text{ such that } y = f(x)\}.$$

Proof. Suppose, $\{B_\alpha\}$ is an open cover for $f(X)$ (in the relative topology). This implies that $B_\alpha = G_\alpha \cap f(X)$ for some $G_\alpha \in \tau_y$. Then the collection of sets $\{f^{-1}(G_\alpha)\}$ forms an open cover of X. From compactness of X there exists a finite set F (i.e. F has finite number of elements) such that the collection $\{f^{-1}(G_\alpha)\}_{\alpha\in F}$ covers X. It is clear that $f(X) = \cup_{\alpha\in F} B_\alpha$. □

Theorem 1.2.3. *(X, τ) is a compact topological set if and only if every universal net in X is convergent.*

Proof. ($\Rightarrow$) Suppose (X, τ) is a compact topological set. Let x_λ be a universal net in X on Λ such that for every $x \in X$ there exists an open set G_x containing x such that x_λ is not eventually in G_x. As x_λ is a universal net it follows that for every $x \in X$, x_λ is eventually in $X \backslash G_x$. It is clear that $\bigcup_{x\in X} G_x$ is an open cover of X. As X is compact there exist elements $x_1, \ldots, x_n$ in X such that $X = \bigcup_{i=1}^n G_{x_n}$. As x_λ is eventually in $X \backslash G_{x_n}$ for all $n = 1, \ldots, n$ it follows that x_λ is eventually in $\bigcap_{i=1}^n X \backslash G_{x_n}$. This implies that x_λ is eventually in $X \backslash \bigcup_{i=1}^n G_{x_n} = \{\}$. Thus we have reached a contradiction and it follows that every universal net in a compact set converges.

($\Leftarrow$) Suppose (X, τ) is not a compact set. Then there exists an open cover $\{A_\lambda\}_{\lambda\in\Lambda}$ such that for every finite subset F of Λ there exists an element x_F in X such that $x_F \in X \backslash \bigcup_{\lambda\in F} A_\lambda$. Let

$$T := \{F : F \subset \Lambda \text{ and } F \text{ is finite}\}.$$

Define an order $\prec$ on T by $F_1 \prec F_2$ if and only if $F_1 \subset F_2$. Then x_F is a net in X on T. Let $(y_\beta)_{\beta\in M}$ with $h : M \to T$ be a universal subnet of the net x_F (we know that such a subnet exists from Theorem 1.2.2). Suppose there exists y_0 such that $y_\beta \to y_0$. Then there exists A_{F_0} with $F_0 \in \Lambda$ such that $y_0 \in A_{F_0}$ and y_β is eventually in A_{F_0}. Note that F_0 is also an element of T. As y_β is a subnet of x_λ we know that there exists $\beta_0 \in M$ such that if $\beta_0 \prec \beta$ then $F_0 \prec h(\beta)$ in T. As y_β is eventually in A_{F_0} we know that there exists a β' such that $\beta' \prec \beta$ implies that $y_\beta \in A_{F_0}$. Let α be the majorant of β_0 and β'. Then $y_\alpha \in A_{F_0}, F_0 \prec h(\alpha)$ and $y_\alpha = x_{h(\alpha)}$. But we know that $x_{h(\alpha)} \in X \backslash \bigcup_{\lambda \in h(\alpha)} A_\lambda$. As $F_0 \subset h(\alpha)$ we have a contradiction. This proves the theorem. □

Corollary 1.2.1. *In a compact topological set (X, τ) every net has a convergent subnet.*

Proof. Follows immediately from Theorem 1.2.2 and Theorem 1.2.3. □

It is not true that in a compact topological set every sequence has a convergent subsequence. We will give examples of this fact later after we have introduced weak $*$ topology. We will use the machinery developed to prove an important theorem; Tychonoff's theorem.

Definition 1.2.19 (Product set, Projection, Product topology). *Suppose $\{X_j\}_{j\in J}$ is a collection of sets with topologies τ_j. The product set denoted by $\Pi_{j\in J}X_j$ is a collection of sets each of which has an element from all the sets X_j. The projection operators π_j are mappings from $\Pi_{j\in J}X_j$ to X_j. If A is in $\Pi_{j\in J}X_j$ then $\pi_j(A)$ called the projection of A onto its jth factor is the element in A which belongs to X_j. The initial topology on $\Pi_{j\in J}X_j$ induced by the functions π_j is called the product topology.*

Example 1.2.2. Let $X_1 = \{x \in R : 1 \le x \le 2\}$ and let $X_2 := \{x \in R : 3 \le x \le 4\}$. Then

$$\Pi_{j=1}^2 X_j := \{(a, b) : a \in X_1 \text{ and } b \in X_2\}.$$

Theorem 1.2.4 (Tychonoff's theorem). *Let $\{X_j : j \in J\}$ be a collection of compact topological sets. The product set $\Pi_{j\in J}X_j$ with the product topology is compact.*

Proof. Let x_λ be a universal net in $\Pi_{j\in J}X_j$. Then, $\pi_j(x_\lambda)$ is a universal net in X_j (see Lemma 1.2.11). As X_j is compact it follows that $\pi_j(x_\lambda)$ is a convergent net (see Theorem 1.2.3). As $\Pi_{j\in J}X_j$ has the product topology it follows that x_λ is convergent in $\Pi_{j\in J}X_j$ (see Lemma 1.2.7). As x_λ is an arbitrary universal net in $\Pi_{j\in J}X_j$ it follows that $\Pi_{j\in J}X_j$ is compact (see Theorem 1.2.3). □

1.3 Metric Topology

The exposition to topology that we have presented untill now is very general and it lacks the structure that most sets we encounter have. The notion of distance between any two elements of a set is an important concept which is made precise by defining an appropriate metric on the set.

Definition 1.3.1 (Metric). *Let X be a nonempty set. A metric on the set X is a map d, from $X \times X$ to the real line which has the following properties.*

1. $d(x, y) \geq 0$ *and* $d(x, y) = 0$ *if and only if* $x = y$;
2. $d(x, y) = d(y, x)$ *(symmetry);*
3. $d(x, y) \leq d(x, z) + d(z, y)$ *(the triangle inequality),*

where x, y and z are elements in X.

The set X with the metric d is denoted as (X, d). A metric on a set X induces a topology on the the set called the metric topology.

Definition 1.3.2 (Metric topology). *Let (X, d) be a metric set. We say that a subset A, of X is open in the metric topology induced by d if for every element a in A there exists an ϵ sufficiently small such that the set $\{x \in X : d(x, a) < \epsilon\}$ is contained in A.*

It can be shown easily that defining open sets in this manner induces a topology on X. We say that (X, d) is a topological metric set when the topology on X is induced by the metric d. Let $A_n(x) := \{y : d(x, y) < \frac{1}{n}\}$ where n is any positive integer and x is any element in X. If N is a neighbourhood of x in the topological metric set (X, d), then it contains a set of the form $\{y : d(x, y) < \epsilon\}$ where $\epsilon > 0$. The set $A_m(x)$ with $\frac{1}{m} < \epsilon$ is contained in $\{y : d(x, y) < \epsilon\}$ and therefore every neighbourhood of x contains a set from the collection $\{A_n(x)\}$. Thus we have shown that any topological metric set satisfies the first axiom of countability. It follows that all convergence results can be established by using the notion of sequences and subsequences and the use of nets and subnets becomes superfluous for topological metric sets.

Lemma 1.3.1. *Let x_0 be any element in a topological metric set (X, d). Let*

$$B_\epsilon(x_0) := \{y : d(y, x_0) \leq \epsilon\},$$

where ϵ is an arbitrary non-negative number. Then, B_ϵ is closed. Also, the set

$$G_\epsilon(x_0) := \{y : d(y, x_0) < \epsilon\},$$

is open.

Proof. If $x \in X \backslash B_\epsilon$ then $d(x, x_0) > \epsilon$. Consider the set $A_x := \{y : d(x, y) < a\}$, where $a = (d(x, x_0) - \epsilon)/2$. If $y \in A_x$ then $d(y, x_0) \geq d(x, x_0) - d(y, x) > d(x, x_0) - a > \frac{d(x, x_0) + \epsilon}{2} > \epsilon$. Therefore, $A_x \subset X \backslash B_\epsilon$. Its clear that $X \backslash B_\epsilon = \bigcup_{x \in X \backslash B_\epsilon} A_x$. As A_x is open for all x in $X \backslash B_\epsilon$ it follows that $X \backslash B_\epsilon$ is open and therefore B_ϵ is closed. We leave the proof of the second assertion to the reader. □

We now summarize convergence results for topological metric sets.

Lemma 1.3.2. *The following assertions are true for a topological metric sets.*

1. *Let x_n be a sequence in a metric set (X, d) and let $x_0 \in X$. $x_n \to x_0$ in the topology induced by d, if and only if for every real $\epsilon > 0$ there exists an integer m such that $n > m$ implies that $d(x_n, x_0) < \epsilon$.*
2. *A function $f : (X, d_X) \to (Y, d_Y)$ is continuous at x_0 if and only if for every sequence x_n which converges to x_0 in (X, d_X), $f(x_n) \to f(x_0)$ in (Y, d_Y).*
3. *A function $f : (X, d_X) \to (Y, d_Y)$ is continuous if and only if for any sequence $\{x_n\}$ convergent in (X, d_X), $f(x_n)$ converges to $f(\lim x_n)$ in (Y, d_Y).*

Proof. (1. $\Rightarrow$) Suppose that the sequence x_n is eventually in every neighbourhood of x_0. Given any $\epsilon > 0$ consider the set $G := \{y : d(y, x_0) < \epsilon\}$. Because this set is a neighbourhood of x_0 it follows that x_n is eventually in G which implies that there exists an integer m such that if $m < n$ then $x_n \in G$. Therefore, for every real $\epsilon > 0$ there exists an integer m such that $n > m$ implies that $d(x_n, x_0) < \epsilon$.

(1. $\Leftarrow$) Proving the other direction is similar and we leave it to the reader. We also leave the proofs of assertions 2 and 3 to the reader. □

The next lemma is an important characterization of compactness for a topological metric space. The proof of the following lemma is left to the reader.

Lemma 1.3.3. *For a topological metric set (X, d) the following statements are equivalent.*

1. *X is compact.*
2. *Every sequence has a convergent subsequence.*

Thus unlike general topological sets, in metric sets compactness is completely characterized in terms of sequences.

Definition 1.3.3 (Cauchy sequence). *A sequence $\{x_n\}$ in a topological metric set (X, d) is a cauchy sequence if for every positive number ϵ, there exists an integer N such that if $n, m > N$ then $d(x_n, x_m) < \epsilon$.*

Note that every convergent sequence is Cauchy. However, it is not true that in a topological metric set all Cauchy sequences are convergent.

Example 1.3.1. Let $X := (0, 1] := \{x : 0 < x \leq 1\}$ with the metric defined by $d(x, y) = |x - y|$ (which is the absolute value of $x - y$). Indeed, the sequence $\frac{1}{n}$ is cauchy but is not convergent.

Definition 1.3.4 (Completeness). *A topological metric set is said to be complete if all Cauchy sequences in it converge.*

Definition 1.3.5 (Isometric maps). *A map $f : (X, d_x) \to (Y, d_y)$ is isometric if*

$$d_x(x_1, x_2) = d_y(f(x_1), f(x_2)),$$

for all elements x_1 and x_2 in X.

1.4 Normed Linear (Vector) Spaces

Linear spaces (often called vector spaces) are sets with the operations of vector addition and scalar multiplication that satisfy a set of properties which are given in the following definition.

Definition 1.4.1 (Vector space). *Let X be a nonempty set and let '+', called the addition operation associate two elements x and y in X with another element $x + y$ in X which satisfies the following properties.*

1. $x + y = y + x$ *for vectors x and y in X.*
2. $x + (y + z) = (x + y) + z$ *for vectors x, y and z in X.*
3. *There exists in X a unique element denoted by 0 and called the zero element, or the origin, such that $x + 0 = x$ for all x in X.*
4. *For every element x in X, there exists a unique element $-x$, called the negative of x, such that $x + (-x) = 0$.*

We will refer to the system of real numbers or to the complex numbers as the scalars. The scalar multiplication operation associates a scalar α and an element x in X with another element αx, in X in such a way that

1. $\alpha(x + y) = \alpha x + \alpha y$.
2. $(\alpha + \beta)x = \alpha x + \beta x$ *for vectors x and scalars α and β.*
3. $(\alpha\beta)x = \alpha(\beta x)$.
4. $\alpha x = x$ *if the scalar $\alpha = 1$.*

The set X with the operations of vector addition and scalar multiplication as defined above is called a vector space or a linear space.

The term 'space' is used to mean a vector space. It should be clear from the above definition that the set of scalars admit two operations; one is the the scalar multiplication and the other is scalar addition. The underlying set of scalars is the real number system unless mentioned otherwise.

Definition 1.4.2 (Subspace). *A subset A, of a vector space X, with $\alpha x + \beta y \in A$ for all scalars α and β and vectors $x \in A$ and $y \in A$ is a subspace of X.*

Note that a subspace of a vector space is also a vector space.

Definition 1.4.3 (Normed vector space). *A normed linear space is a vector space X with a function $||\cdot|| : X \to R$ defined such that*

1. $||x|| \geq 0$ *and* $||x|| = 0$ *if and only if* $x = 0$.
2. $||\alpha x|| = |\alpha|\ ||x||$ *for any scalar α and vector x in X.*
3. $||x + y|| \leq ||x|| + ||y||$.

It is clear that a norm on a vector space X induces a metric on X defined by $d(x, y) := ||x - y||$ for elements x and y in X. Therefore, a norm also induces a topology (called the *norm topology*) which is the metric topology with the metric defined as above. The normed topological space X with a norm $||\cdot||$ is denoted by $(X, ||\cdot||)$. Also, note that because a normed space is a metric space sequences suffice to describe convergence.

Example 1.4.1. Consider the set R^n which is defined as

$$R^n := \{(x_1, x_2, \ldots, x_n) : x_i \in R \text{ for } i = 1, \ldots, n\}.$$

R^n is a vector space with the real numbers as the scalars. Many different norms can be defined on R^n. An important class of norms on R^n is defined by the p norm which is given by

$$|x|_p := (\sum_{i=1}^{n} |x_i|^p)^{\frac{1}{p}}$$

where $x = (x_1, x_2, \ldots, x_n)$ and p is an integer such that $1 \leq p < \infty$. The two norm ($p = 2$) and the one norm ($p = 1$) are of particular interest. Another important norm on R^n is the ∞ norm which is defined as

$$|x|_\infty := \max_{1 \leq i \leq n} |x_i|.$$

Example 1.4.2 (Product normed spaces). Let $(X, ||\cdot||_X)$ and $(Y, ||\cdot||_Y)$ be normed vector spaces. Let $X \times Y := \{(x, y) : x \in X \text{ and } y \in Y\}$. For (x, y) in $X \times Y$ and $\alpha \in R$, $\alpha(x, y) \in X \times Y$ is defined as $(\alpha x, \alpha y)$. Also, for elements (x_1, y_1) and (x_2, y_2) in $X \times Y$, $(x_1, y_1) + (x_2, y_2)$ is defined as $(x_1 + x_2, y_1 + y_2)$. With these operations $X \times Y$ is a vector space. For (x, y) in $X \times Y$ let $||\cdot|| \to R$ be defined by

$$||(x, y)|| := ||x||_X + ||y||_Y.$$

Then it can be shown that $||\cdot||$ is a norm on $X \times Y$. The normed vector space $(X \times Y, ||\cdot||)$ is called the product space of X and Y and $||\cdot||$ is called the product norm.

Lemma 1.4.1. *In a normed topological vector space $(X, ||\cdot||)$, the function $||\cdot|| : X \to R$ is a continuous function.*

Proof. Let $\{x_n\}$ be any sequence in X. Note that $||x_n|| = ||x_n - x_0 + x_0|| \leq ||x_n - x_0|| + ||x_0||$ which implies that $||x_n|| - ||x_0|| \leq ||x_n - x_0||$. Using the inequality $||x_0|| \leq ||x_n - x_0|| + ||x_n||$ it follows similarly that $||x_0|| - ||x_n|| \leq ||x_n - x_0||$. Therefore, $|\ ||x_n|| - ||x_0||\ | \leq ||x_n - x_0||$. This implies that if x_n converges to x_0 in X then $||x_n||$ converges to $||x_0||$ in R. Therefore from Lemma 1.3.2 it follows that $||\cdot||$ is a continuous function. □

We have given an example in the previous section to illustrate that not all Cauchy sequences converge. However the following lemma shows that every Cauchy sequence is bounded.

Lemma 1.4.2. *Let $\{x_n\}$ be a Cauchy sequence in a normed linear space $(X, ||\cdot||)$. Then there exists a positive real number M such that $||x_n|| \leq M$ for all n.*

Proof. As $\{x_n\}$ is Cauchy there exists an integer N such that $n, m \geq N$ implies that $||x_n - x_m|| \leq 1$. Choose any $n > N$. Then $||x_n|| = ||x_n - x_N + x_N|| \leq ||x_n - x_N|| + ||x_N|| \leq 1 + ||x_N||$. This proves the lemma. □

Of particular importance to optimization are Banach spaces.

Definition 1.4.4 (Banach spaces). *A complete normed vector space is a Banach space.*

The *a priori* knowledge of the existence of a limit point for a Cauchy sequence is helpful in optimization problems where construction of Cauchy sequences is natural.

Definition 1.4.5 (Linear independence, Basis, Dimension).

In a vector space X, vectors $x_1, x_2, \ldots, x_n$ are linearly independent if for any set of scalars $\alpha_1, \alpha_2, \ldots, \alpha_n$

$$\alpha_1 x_1 + \alpha_2 x_2 + \ldots + \alpha_n x_n = 0$$

implies that $\alpha_i = 0$ for all $i = 1, \ldots, n$. If vectors $x_1, \ldots, x_n$ are not linearly independent then they are called dependent.

A set of linearly independent vectors $x_1, x_2, \ldots, x_n$ is a basis for the vector space X if for any vector y in X there exists scalars $\alpha_1, \alpha_2, \ldots, \alpha_n$ such that $y = \sum_{i=1}^{n} \alpha_i x_i$.

If there exist n vectors which form a basis for a vector space X then X has dimension n and is said to be finite dimensional. *If no finite number of vectors can form a basis for the vector space X then X is infinite dimensional.*

Finite dimensional normed spaces have structure which might be missing from infinite dimensional spaces. Now we study linear operators from one normed space to another. As we will see in the coming chapters linear functions play an extremely important role in optimization theory.

Definition 1.4.6 (Linear map, Affine linear map). *A map T, from a vector space X to a vector space Y is linear if for any scalars α and β and vectors x_1 and x_2, $T(\alpha x_1 + \beta x_2) = \alpha T(x_1) + \beta T(x_2)$.*

A map S, from a vector space X to a vector space Y is affine linear if the map $S - S(0)$ is linear.

Example 1.4.3. Let $A : R^2 \to R^2$ be defined by

$$A(x) := \begin{pmatrix} a_{11} & a_{12} \\ a_{21} & a_{22} \end{pmatrix} \begin{pmatrix} x_1 \\ x_2 \end{pmatrix},$$

where $x := \begin{pmatrix} x_1 \\ x_2 \end{pmatrix}$. Then the map A is linear. The map $H : R^2 \to R^2$ with $H(x) = A(x) + b$ where $b \in R^2$ is an affine linear map.

Theorem 1.4.1. *Let $(X, ||\cdot||_X)$ and $(Y, ||\cdot||_Y)$ be normed spaces. For a linear map $T : (X, ||\cdot||_X) \to (Y, ||\cdot||_Y)$ the following conditions are equivalent.*

1. *T is continuous.*
2. *T is continuous at the origin i.e. if the sequence $x_n \to 0$ in $(X, ||\cdot||_X)$ then $T(x_n) \to 0$ in $(Y, ||\cdot||_Y)$.*
3. *There exists a real number $K \geq 0$ such that $||T(x)||_Y \leq K||x||_X$.*

Proof. $(1 \Rightarrow 2)$ Follows immediately because every continuous map is continuous at every point.

$(2 \Rightarrow 1)$ Suppose for any sequence $\{x_n\}$ in X which converges to zero, $T(x_n)$ converges to zero. Let $\{z_n\}$ be a sequence which converges to z_0 in X. It follows clearly that the sequence $\{z_n - z_0\}$ converges to zero in X and therefore $T(z_n - z_0)$ converges to zero. From linearity of the map T it follows that $T(z_n) - T(z_0)$ converges to zero which implies that $T(z_n)$ converges to $T(z_0)$. Thus we have shown that if z_n converges to z_0 in X then $T(z_n)$ converges to $T(z_0)$ in Y.

$(2 \Rightarrow 3)$ Suppose for every positive integer n, there exists an element x_n in X such that $||T(x_n)||_Y > n||x_n||_X$. Let $z_n := \frac{x_n}{n||x_n||_X}$. Then $||z_n||_X = \frac{1}{n}$ and therefore z_n converges to 0 in X. However, $||T(z_n)|| > 1$ for all n which implies that $T(z_n)$ does not converge to zero. This contradicts the fact that $z_n \to 0$.

$(3 \Rightarrow 2)$ Suppose there exists a real number K such that $||T(x)||_Y \leq K||x||_X$ for all x in X. Let x_n be a sequence in X which converges to zero. Then, because $||T(x_n)||_Y \leq K||x_n||_X$ it follows that $T(x_n)$ converges to zero in Y. □

A map f, from a normed vector space X to a normed vector space Y is said to be a *bounded map* if there exists a real number K such that $||f(x)|| \leq K||x||$. Thus for a linear map the adjectives continuous and bounded can be used interchangeably. For any map f which maps a normed vector space $(X, ||\cdot||_X)$ to another normed vector space $(Y, ||\cdot||_Y)$ we define $||f|| := \sup\{||f(x)||_Y : ||x||_X \leq 1\}$.

1.5 Finite Dimensional Spaces

Finite dimensional spaces enjoy a number of properties which do not hold for general infinite dimensional spaces. Here, we show that any normed finite dimensional space with dimension n is essentially similar to R^n with any norm defined.

Lemma 1.5.1. *Let I be a closed bounded interval in R (that is $I = [a, b]$ with a and b in R). Then I is compact.*

Proof. Let $\{A_\lambda\}_{\lambda \in \Lambda}$ be a open covering of I. Let
$$r = \sup\{t : a \leq t \leq b+1, \text{ such that } [a, t] \text{ has a finite subcover from } \{A_\lambda\}\}.$$

Suppose $r \leq b$. Then there exists $\lambda_0 \in \Lambda$ such that $r \in A_{\lambda_0}$. As A_{λ_0} is open there exists an $\epsilon > 0$ such that $[r-\epsilon, r+\epsilon] \subset A_{\lambda_0}$. From the definition of r it follows that $[a, r-\epsilon]$ admits a finite subcover from $\{A_\lambda\}$. This subcover together with A_{λ_0} forms a finite subcover of $[a, r+\epsilon]$. This contradicts the definition of r. Thus $r > 1$. This proves the lemma. □

Now we prove the well known Hiene-Borel theorem.

Theorem 1.5.1 (Hiene-Borel). *Every closed bounded subset C, of $(R^n, |\cdot|_1)$ is compact.*

Proof. Because C is bounded there exist bounded closed intervals I_k for $k = 1, \ldots, n$ in R such that $C \subset \Pi_{k=1}^n I_k$. As I_k are compact (see Lemma 1.5.1), it follows from Theorem 1.2.4 (Tychonoff's theorem) that $\Pi_{k=1}^n I_k$ is compact. As C is a closed subset of $\Pi_{k=1}^n I_k$ it follows from Lemma 1.2.12 that C is compact. □

Lemma 1.5.2. *Every linear map from $(R^n, |\cdot|_1)$ to any normed space $(X, ||\cdot||_X)$ is continuous.*

Proof. Let e_i be the natural basis for R^n where e_i is the n-tuple with 1 in the ith place and zeros elsewhere. Suppose, $x_k \to x_0$ in $(R^n, |\cdot|_1)$ that is, $|x_k - x_0|_1 \to 0$. If $x_k = \sum_{i=1}^n \alpha_i^k e_i$ and $x_0 = \sum_{i=1}^n \alpha_i^0 e_i$ then this implies that $\alpha_i^k \to \alpha_i^0$ for all $i = 1, \ldots, n$. Now, if $T : (R^n, |\cdot|_1) \to (X, ||\cdot||_X)$ is linear then $||T(x_k) - T(x_0)|| = ||T(x_k - x_0)|| = ||T(\sum_{i=0}^n (\alpha_i^k - \alpha_i^0)e_i)|| \leq \max_{1 \leq i \leq n} |\alpha_i^k - \alpha_i^0| \sum_{i=0}^n ||T(e_i)||$. As $|\alpha_i^k - \alpha_i^0| \to 0$ for all $i = 1, \ldots, n$, it follows that $T(x_k)$ converges to $T(x_0)$. Therefore, T is continuous. □

The proof above is essentially the consequence of the fact that in $(R^n, |\cdot|_1)$ convergence is equivalent to componentwise convergence. In the next lemma we show that for any finite dimensional normed space there exists a map between it and $(R^n, |\cdot|_1)$ which is continuous and its inverse (which exists) is also continuous.

Lemma 1.5.3. *Let $(X, ||\cdot||)$ be an n dimensional normed vector space with the vectors $\{x_1, \ldots, x_n\}$ forming a basis for X. For every $x \in X$ there exists a unique set of n scalars α_i, with $i = 1, \ldots, n$ such that $x = \sum_{i=1}^n \alpha_i x_i$. If T denotes the map from $(X, ||\cdot||)$ to $(R^n, |\cdot|_1)$ where $|\cdot|_1$ is the one norm on R^n and $T(x) := (\alpha_1, \ldots, \alpha_n)$ for $x = \sum_{i=1}^n \alpha_i x_i$, then T is a linear homeomorphism.*

Proof. Suppose $x = \sum_{i=1}^n \alpha_i x_i = \sum_{i=1}^n \beta_i x_i$. then it follows that $\sum_{i=1}^n (\alpha_i - \beta_i) x_i = 0$. As x_i are independent it follows that $\alpha_i = \beta_i$ for $i = 1, \ldots, n$. Therefore, there exists a unique set of scalars α_i, such that $x = \sum_{i=1}^n \alpha_i x_i$. This implies that T is a well defined function.

It is clear that T is one-to-one, onto and linear (the proof is left to the reader). As T is one-to-one and onto $T^{-1}(x)$ consists only of a single element from X and therefore it is a linear map from $(R^n, |\cdot|_1)$ to $(X, ||\cdot||)$. From Lemma 1.5.2 it follows that T^{-1} is a continuous map.

Suppose T is not continuous. Then from Theorem 1.4.1, T is not continuous at zero. Therefore, there exists a sequence $\{x_k\}$ such that $x_k \to 0$ in $(X, ||\cdot||)$ but $T(x_k) \not\to T(x_0)$ in $(R^n, |\cdot|_1)$. This implies that there exists an $\epsilon > 0$ such that for every integer j there exists x_{k_j} with $k_j \geq j$ and $|T(x_{k_j})|_1 \geq \epsilon$. Let $y_j := \frac{x_{k_j}}{|T(x_{k_j})|_1}$. Then, $|T(y_j)|_1 = 1$ and $y_j \to 0$ in $(X, ||\cdot||)$. $T(y_j)$ is a sequence inside the compact set $B := \{a : a \in R^n, |a|_1 \leq 1\}$ and therefore from Theorem 1.5.1 (Hiene-Borel theorem) and Lemma 1.3.3 it follows that there exists a subsequence of $T(y_j)$ which converges to some element p in R^n. Without loss of generality we assume that the sequence $T(y_j)$ is convergent. As the norm is a continuous function it follows that $|p|_1 = 1$. Because, T^{-1} is a continuous function and $T(y_j) \to p$ it follows that $T^{-1}(T(y_j)) \to T^{-1}(p)$ and therefore $y_j \to T^{-1}(p)$. But $y_j \to 0$ and therefore $T^{-1}p = 0$ which implies that $p = 0$. This is a contradiction to the fact that $|p|_1 = 1$. Therefore, T is continuous. □

Thus we have shown that there exists a homeomorphism between any finite dimensional normed vector space with dimension n and $(R^n, |\cdot|_1)$. It follows that there exists a homeomorphism between any two normed finite dimensional spaces which have the same dimension.

Corollary 1.5.1. *Let $(X, ||\cdot||_a)$ and $(X, ||\cdot||_b)$ be the same n dimensional vector space X, with two different norms defined. Then there exists constants m and M such that*

$$m||x||_a \leq ||x||_b \leq M||x||_a.$$

Proof. Let the basis for the vector space X be given by $\{x_1, \ldots, x_n\}$. From Lemma 1.5.3 it follows that there exists a map $T : (X, ||\cdot||_a) \to (R^n, |\cdot|_1)$ such that if $x = \sum_{i=1}^{n} \alpha_i x_i$ then $T(x) = (\alpha_1, \ldots, \alpha_n)$ and both T and T^{-1} are continuous linear maps (the fact that T^{-1} is linear is left for the reader to prove). From Theorem 1.4.1 it follows that there exist constants K_1 and K_2 such that $|T(x)|_1 \leq K_1||x||_a$ and $||x||_a = ||T^{-1}(T(x)||_a \leq K_2|T(x)|_1$. Therefore, there exist constants K_1 and K_2 such that $\frac{1}{K_1}|T(x)|_1 \leq ||x||_a \leq K_2|T(x)|_1$. Similarly, there exist constants N_1 and N_2 such that $\frac{1}{N_1}|T(x)|_1 \leq ||x||_b \leq N_2|T(x)|_1$. The corollary follows easily from these inequalities. □

Corollary 1.5.2. *Every finite dimensional normed vector space is complete and every finite dimensional subspace of a normed vector space is closed in the relative topology.*

Proof. Let $(X, ||\cdot||)$ be a finite dimensional normed vector space. Let T be the homeomorphism between $(X, ||\cdot||)$ and $(R^n, |\cdot|_1)$ (as given in Lemma 1.5.3). Suppose x_k is a Cauchy sequence in $(X, ||\cdot||)$. As T is bounded there exists a constant M such that $|T(x)|_1 \leq M||x||$. This implies that $|T(x_k) - T(x_m)|_1 = |T(x_k - x_m)|_1 \leq M||x_k - x_m||$. As the sequence $\{x_k\}$ is Cauchy it follows that $\{T(x_k)\}$ is Cauchy. $(R^n, |\cdot|_1)$ is a complete space (left to the reader to prove) and therefore $T(x_k)$ converges to some element $q \in R^n$. As T^{-1} is continuous it follows that $T^{-1}(T(x_k)) \to T^{-1}(q)$. Therefore x_k is convergent in X. Therefore, $(X, ||\cdot||)$ is a complete space. The proof of the second assertion, which is a consequence of the first assertion is left to the reader. □

The following theorem says that the only vector spaces in which all norm closed and bounded sets are compact are finite dimensional.

Theorem 1.5.2. *In a normed vector space $(X, ||\cdot||)$, the set $B := \{x : x \in X$ and $||x|| \leq 1\}$ is compact if and only if X is finite dimensional.*

Proof. ($\Leftarrow$) Suppose X is a finite dimensional vector space with dimension n. From Lemma 1.5.3 we know that there exists a linear homeomorphism, $T : (X, ||\cdot||) \to (R^n, |\cdot|_1)$. Therefore there exists a constant M such that $|T(x)|_1 \leq M||x||$ for all $x \in X$. Therefore, it follows that $T(B)$ is a bounded set in $(R^n, |\cdot|_1)$. From Lemma 1.3.1 we know that B is closed. As T is continuous it follows from Lemma 1.2.5 that $T(B)$ is closed. Therefore, $T(B)$ is compact in $(R^n, |\cdot|_1)$ (see Theorem 1.5.1; Hiene-Borel theorem) . As T^{-1} is continuous it follows that $B = T^{-1}(T(B))$ is compact (Lemma 1.2.14).

($\Rightarrow$) Let C_1 and C_2 be two subsets of a vector space X. Then we define the sum of the two subsets as

$$C_1 + C_2 := \{z : \text{ there exists } x \in C_1 \text{ and } y \in C_2 \text{ such that } z = x + y.\}$$

Also, we define for a scalar α and a subset C_1 of the vector space X,

$$\alpha C_1 := \{z : \text{ there exists } x \in C_1 \text{ such that } z = \alpha x\}.$$

Let $B(x, \frac{1}{2}) := \{y : ||x - y|| < \frac{1}{2}\}$ for any x in X. Then $\cup_{x \in B} B(x, \frac{1}{2})$ is an open cover of B. As B is compact there exist vectors $x_1, x_2 \ldots, x_n$ in B such that $B = \cup_{i=1}^n B(x_i, \frac{1}{2})$. Note that $B(x_i, \frac{1}{2}) = \{x_i\} + \frac{1}{2}B$. Therefore, $B = \cup_{i=1}^n (\{x_i\} + \frac{1}{2}B)$. Let $Y = \text{span}\{x_1, x_2, \ldots, x_n\}$. Then it follows that $B = Y + \frac{1}{2}B \subset Y + \frac{1}{2}(Y + \frac{1}{2}B) \subset Y + \frac{1}{2}(Y + \frac{1}{2}(Y + \frac{1}{2}B))$. Continuing in this manner we can establish that $B \subset Y + \frac{1}{2^m}B$ for any positive integer m. For every $x \neq 0$, $\frac{x}{||x||} \in B$ and therefore given $x \in X$ and $m > 0$ there exists $y_m \in Y$ and $b_m \in B$ such that $y_m + 2^{-m}b_m = \frac{x}{||x||}$. Because Y is a vector space for every $x \in X$ and $m > 0$ there exists $y_m \in Y$ and $b_m \in B$ such that $y_m + 2^{-m}b_m = x$. This implies that for any $x \in X$ and $m > 0$ there exists $y_m \in Y$, $b_m \in B$ such that $||y_m - x|| = 2^{-m}||b_m|| \leq 2^{-m}$. Therefore, $X = Y^-$. But Y^- is Y itself because Y is finite dimensional (see Corollary 1.5.2). Therefore, X is finite dimensional. □

This Lemma indicates the scarceness of compact sets in the norm topology. As we will see in the next section compactness is essential in optimization and this will lead us to define less restrictive topologies in the next chapter.

1.6 Extrema of Real Valued Functions

In this section we provide characterizations for functions which allow for extrema to exist. It is shown that compactness of sets and continuity properties of functions play an important role for extrema to exist.

Definition 1.6.1 (Local extrema). *Let $(X, \|\cdot\|_X)$ be a normed vector space and let $f : D \to R$ be a real valued function defined on a subset D of X. An element x_0 in X is a local minimum if there exists a neighbourhood N of x_0 such that for all $x \in N \cap D$, $f(x_0) \leq f(x)$. x_0 is a strict minimum if for all $x \in N \cap D$, $f(x_0) < f(x)$. Local maxima are defined analogously. Local extrema refers to either local minima or local maxima.*

Theorem 1.6.1. *If (X, τ) is a topological compact set and $f : (X, \tau) \to R$ is a real valued continuous function then there exists elements x_0 and x_1 in X such that $f(x_0) \leq f(y)$ and $f(x_1) \geq f(y)$ for all $y \in X$.*

Proof. Let $\mu := \inf\{f(x) : x \in X\}$. Then from the definition of infimum for every positive integer n there exists an element x_n such that $f(x_n) \leq \mu + \frac{1}{n}$. Note that as x_n is a sequence in a compact set, from Corollary 1.2.1 it follows that there exists a subnet y_λ of x_n in X on a set M such that $y_\lambda \to x_0$ for some x_0 in X. As f is continuous it follows that $f(y_\lambda) \to f(y_0)$.

As y_λ is a subnet of x_n in X on M there exists a map $h : M \to I$ where I is the set of positive integers such that $x_{h(\lambda)} = y_\lambda$. Also, for every n in I there exists a λ_n such that $\lambda_n \prec \lambda$ implies that $n \leq h(\lambda)$. Now, $f(y_\lambda) = f(x_{h(\lambda)}) \leq \mu + \frac{1}{h(\lambda)}$. Given any $\epsilon > 0$ choose $N \in I$ such that $\frac{1}{N} < \frac{\epsilon}{2}$. If $\lambda_N \prec \lambda$ then $N \leq h(\lambda)$ and therefore $f(y_\lambda) \leq \mu + \frac{1}{N} \leq \mu + \frac{\epsilon}{2}$. As $f(y_\lambda) \to f(x_0)$ we know that given $\epsilon > 0$ there exists λ_ϵ such that if $\lambda_\epsilon \prec \lambda$ then $|f(y_\lambda) - f(x_0)| < \frac{\epsilon}{2}$. Let λ_0 be the majorant of λ_N and λ_ϵ. If $\lambda_0 \prec \lambda$ then $f(x_0) \leq \mu + \epsilon$ and because ϵ is an arbitrary positive real number it follows that $f(x_0) \leq \mu$. As $x_0 \in X$ it follows that $f(x_0) = \mu$. Thus we have established that x_0 attains the infimum.
The proof for the existence of x_1 is similar and is left to the reader. □

Definition 1.6.2 (Lower semicontinuity, Upper semicontinuity). *Let (X, τ) be a topological set and let $f : (X, \tau) \to R$ be a real valued function such that every set of the form $\{x : f(x) > \alpha\}$ for α real is open. Then f is a lower semicontinuous function. f is upper semicontinuous if $-f$ is lower semicontinuous.*

For Λ a directed set and r_λ a net in R on Λ we use the notation $\liminf r_\lambda$ to represent

$$\liminf r_\lambda := \sup_{\lambda_0 \in \Lambda} \left(\inf_{\lambda_0 \prec \lambda} r_\lambda \right),$$

and we use $\limsup r_\lambda$ to represent

$$\limsup r_\lambda := \inf_{\lambda_0 \in \Lambda} \left(\sup_{\lambda_0 \prec \lambda} r_\lambda \right).$$

Lemma 1.6.1. *Let (X, τ) be a topological compact set and $f : (X, \tau) \to R$. f is lower semicontinuous if and only if*

$$f(\lim x_\lambda) \le \liminf f(x_\lambda),$$

for every convergent net x_λ. f is upper semicontinuous if and only if

$$f(\lim x_\lambda) \ge \limsup f(x_\lambda),$$

for every convergent net x_λ.

Proof. ($\Rightarrow$) Suppose f is a lower semicontinuous function and suppose x_λ is a convergent net with $\lim x_\lambda = x_0$. Choose any real number t such that $t < f(x_0)$. As f is lower semicontinuous it follows that the set $\{x : f(x) > t\}$ is in τ. Note that x_0 is in this set and therefore $\{x : f(x) > t\}$ is a neighbourhood of x_0. As $x_\lambda \to x_0$ it follows that x_λ is eventually in this set. Therefore, there exists a λ_1 such that if $\lambda_1 \prec \lambda$ then $f(x_\lambda) > t$ which implies that $\inf_{\lambda_1 \prec \lambda} f(x_\lambda) \ge t$. Therefore, $\liminf f(x_\lambda) = \sup_{\lambda_0 \in \Lambda} \inf_{\lambda_0 \prec \Lambda} f(x_\lambda) \ge t$. This is true for any $t < f(x_0)$ and therefore $\liminf f(x_\lambda) \ge f(x_0)$.

($\Leftarrow$) Suppose, for every convergent net x_λ, $f(\lim x_\lambda) \le \liminf f(x_\lambda)$. Consider any set $F := \{x : f(x) \le t\}$ where $t \in R$. Suppose, $x_0 \in F^-$. Then from Lemma 1.2.3 it follows that there exists a net x_λ in X on a directed set Λ such that $x_\lambda \to x_0$. From the assumption we have $f(x_0) \le \liminf f(x_\lambda) \le t$. Therefore, $x_0 \in F$ which implies that F is closed. It follows that $\{x : f(x) > t\} = X \backslash F$ is open.

Thus we have shown that f is lower semicontinuous if and only if $f(\lim x_\lambda) \le \liminf f(x_\lambda)$. The rest of the lemma is left as an exercise. □

Corollary 1.6.1. *If (X, τ) is a topological compact set and $f : (X, \tau) \to R$ is a real valued lower semicontinuous function then there exists an element x_0 in X such that $f(x_0) \le f(y)$ for all $y \in X$. Similarly, if f is upper semicontinuous then there exist an element x_1 in X such that $f(x_1) \ge f(y)$ for all $y \in X$.*

Proof. Follows from Lemma 1.6.1 and arguments similar to one used in proving Theorem 1.6.1. □

It is clear that the topology of a set is vital in determining whether extrema for a function exist or not. In many cases the function is a measure of a physical quantity which needs to be minimized or maximized on the given set. We have seen in the previous section that the norm topology is particularly restrictive for infinite dimensional spaces because of the dearth of compact sets in this topology (a norm bounded ball is not compact in the norm topology; see Theorem 1.5.2). Therefore, for infinite dimensional spaces it is worthwhile to study relevant topologies other than the norm topology. We do this in the next chapter.

The fact that the derivative of a real valued function $f : R \to R$ vanishes when it has a local maxima or a local minima is a classical result. Now, we generalize this result.

Definition 1.6.3 (Gateaux derivative). *Let X be a vector space with an open subset D and let $(Y, ||\cdot||_Y)$ be a normed vector space with a map, $f : D \to Y$ defined. For an element $x \in X$ and $h \in X$, f is said to be gateaux differentiable at x with increment h if there exists an element $f_h(x) \in Y$ such that*

$$||\frac{f(x+\alpha h) - f(x) - \alpha f_h(x)}{\alpha}||_Y \to 0 \text{ as } \alpha \to 0.$$

$f_h(x)$ is called the gateaux derivative of f at x in the direction h. If f is gateaux differentiable at x with all increments $h \in X$ then f is said to be gateaux differentiable at x. If f is gateaux differentiable at all $x \in X$ then f is gateaux differentiable.

Note that for a differentiable function $f : R \to R$ the notion of the gateaux derivative and the ordinary derivative are the same.

Theorem 1.6.2. *Let $f : (X, ||\cdot||_X) \to R$ be a real valued gateaux differentiable function on a normed vector space $(X, ||\cdot||_X)$. An element x_0 in X is a local extrema only if $f_h(x_0) = 0$ for all $h \in X$.*

Proof. Suppose at x_0 there is a local minima. Then there exists an $\epsilon > 0$ such that $||z||_X \leq \epsilon$ implies that $f(x_0 + z) - f(x_0) \geq 0$. Therefore, if $\alpha > 0$ and $||\alpha h||_X \leq \epsilon$ then $\frac{f(x_0 + \alpha h) - f(x_0)}{\alpha} \geq 0$. Letting $\alpha \to 0$ while keeping α positive we see that $f_h(x) \geq 0$. Similarly, note that if $\alpha < 0$ and $||\alpha h||_X \leq \epsilon$ then $\frac{f(x_0 + \alpha h) - f(x_0)}{\alpha} \leq 0$. Letting $\alpha \to 0$ while keeping α negative we see that $f_h(x_0) \leq 0$. Thus $f_h(x_0) = 0$.

Similarly, it can be shown that if at x_0 there is a local maximum then for all $h \in X$, $f_h(x_0) = 0$. □

2. Functions on Vector Spaces

One of the most important results in optimization theory in vector spaces is the Hahn-Banach theorem which admits many versions. The extension and the geometric versions of the Hahn-Banach theorem are the most important for our development. In this chapter we develop the extension form and leave the geometric form for the next chapter. Very intimately related to the Hahn-Banach theorem is the study of sublinear functions. Sublinear functions also provide the basis for convex analysis. We study sublinear functions in the first section of this chapter.

In the previous chapter, the importance of compactness of a set and the continuity properties of a function were elucidated. It was also shown that in infinite dimensional spaces there is a dearth of compact sets in the norm topology (see Theorem 1.5.2). This neccessitates the search for other topologies where the conditions for compactness are relatively easier to satisfy. This leads us to the study of dual spaces and weak topologies. The study of dual spaces which is essentially the study of bounded linear functions, also lays down the foundation for separation of convex disjoint sets.

We end this chapter with a study of ℓ_p spaces which serve as an example for the various concepts presented. These spaces are also important in the formulation of discrete-time robust control problems.

2.1 Sublinear Functions

The various versions of the Hahn-Banach theorem, the results on linear functions and the weak topologies are related to sublinear functions. We study sublinear functions with the aim of establishing the Hahn-Banach theorem. We call a function $f : X \to R \cup \{\infty\} \cup \{-\infty\}$ *real valued* on the set X if $f(x) \in R$ for all $x \in X$.

Definition 2.1.1 (Sublinear functions). *Let X be a vector space. We say $f : X \to R$ is a sublinear function if it satisfies the following properties.*

1. *For elements x and y in X, $f(x+y) \leq f(x)+f(y)$. Any function satisfying this property is called a subadditive function.*
2. *For scalar $\alpha \geq 0$ and x in X, $f(\alpha x) = \alpha f(x)$. Any function satisfying this property is called a positively homogeneous function.*

Note that for a sublinear function f, $f(0) = 0$.

Lemma 2.1.1. *Let X be a vector space and $S : X \to R$ and $T : X \to R \cup \{\infty\}$ be functions on X such that the following conditions hold.*

1. *For all elements x and y in X, $S(x+y) \leq S(x) + S(y)$ and $T(x+y) \leq T(x) + T(y)$.*
2. *For all element x in X and real numbers $\alpha > 0$, $S(\alpha x) = \alpha S(x)$ and $T(\alpha x) = \alpha T(x)$.*
3. *$S(0) = T(0) = 0$.*

Define

$$U(z) := \inf\{S(z-v) + T(v) : v \in X\}.$$

Also, suppose there exists a $u \in X$ such that $U(u) > -\infty$. Then U is a real valued sublinear function such that for all x in X, $U(x) \leq S(x)$ and $U(x) \leq T(x)$.

Proof. Given elements x and y in X,

$$\begin{aligned} U(x+y) &= \inf\{S(x+y-v) + T(v) : v \in X\} \\ &= \inf\{S(x+y-v_1-v_2) + T(v_1+v_2) : v_1, v_2 \in X\} \\ &\leq \inf\{S(x-v_1) + T(v_1) + S(y-v_2) + T(v_2) : v_1, v_2 \in X\} \\ &= U(x) + U(y). \end{aligned}$$

The second equality is true because $\{v \in X\} = \{v_1 + v_2 \in X\}$ (as X is a vector space), and the inequality is true because of the subadditivity of S and T. This establishes the subadditivity of U.

Suppose α is a positive real number, then,

$$\begin{aligned} U(\alpha x) &= \inf\{S(\alpha x - v) + T(v) : v \in X\} \\ &= \inf\{S(\alpha x - \alpha v) + T(\alpha v) : v \in X\} \\ &= \inf\{\alpha S(x-v) + \alpha T(v) : v \in X\} \\ &= \alpha U(x). \end{aligned}$$

The second equality above is true because $\{v \in X\} = \{\alpha v \in X\}$ (as X is a vector space).

Suppose $x \in X$ then as $0 \in X$ we have $U(x) = \inf\{S(x-v) + T(v) : v \in X\} \leq S(x) + T(0) = S(x)$ and as $x \in X$ we have $U(x) = \inf\{S(x-v) + T(v) : v \in X\} \leq S(0) + T(x) = T(x)$. This implies that $U(x) \leq S(x)$ and $U(x) \leq T(x)$ for all x in X.

Note that $U(x) \leq S(x) < \infty$ for all x in X. Since U satisfies $U(u) > -\infty$ for some $u \in X$, we have that $U(u)$ is real (that is $|U(u)| < \infty$). Therefore, $U(u) = U(u+0) \leq U(u) + U(0)$ which implies that $U(0) \geq 0$. However, $U(0) \leq S(0) = 0$. Hence $U(0) = 0$. Note that for any x in X, $U(x) + U(-x) \geq U(x-x) = 0$. That is, $U(x) \geq -U(-x) \geq -S(-x) > -\infty$. This implies that for all x in X, $U(x) > -\infty$. We have established earlier that $U(x) < \infty$ and thus U is a real valued function. □

Lemma 2.1.2. *Let $S : X \to R$ be a sublinear function on a vector space X, and let x_0 be an element in X. Define for every $w \in X$*

$$U(w) := \inf\{S(w + \alpha x_0) - \alpha S(x_0) : \alpha \geq 0\}.$$

Then U is a real valued sublinear function such that for all $x \in X$, $U(x) \leq S(x)$ and $U(-x_0) \leq -S(x_0)$.

Proof. We assume that $x_0 \neq 0$ (otherwise it follows that for $w \in X$, $U(w) = \inf\{S(w) - \alpha S(0) : \alpha \geq 0\} = S(w)$). Define

$$A := \{y \in X : \text{ there exists } \alpha \in R,\ \alpha \geq 0 \text{ with } y = -\alpha x_0\},$$

and $B = X \backslash A$. Define a function T on X as;

$$\begin{aligned} T(y) &= -\alpha S(x_0) \text{ if } y \in A \text{ with } y = -\alpha x_0 \text{ where } \alpha \geq 0, \\ &= \infty \qquad\quad \text{if } y \in B. \end{aligned}$$

If x and y are elements in A then it follows easily that $T(x+y) = T(x)+T(y)$ and if α is any non-negative real number then $\alpha x \in A$ and $T(\alpha x) = \alpha T(x)$. If $x \in A$ and $z \in B$ then $T(z) = \infty$ and it follows that $T(x + z) \leq T(x) + T(z)$. Also, if $z \in B$ and $\alpha > 0$ then $\alpha z \in B$ and $T(\alpha z) = \infty = \alpha T(z)$. Note that $U(0) := \inf\{S(\alpha x_0) - \alpha S(x_0) : \alpha \geq 0\} = \inf\{\alpha S(x_0) - \alpha S(x_0) : \alpha \geq 0\} = 0$. Also, for any $w \in X$,

$$\begin{aligned} \inf\{S(w - v) + T(v) : v \in X\} &= \inf\{S(w - v) + T(v) : v \in A\} \\ &= \inf\{S(w + \alpha x_0) - \alpha S(x_0) : \alpha \geq 0\} \\ &= U(w). \end{aligned}$$

The second equality is true because $T(v) = \infty$ if $v \notin A$. Thus, S, T and U satisfy the conditions stipulated in Lemma 2.1.1. Therefore, it follows from Lemma 2.1.1 that U is a real valued sublinear function on X such that $U(x) \leq S(x)$ and $U(x) \leq T(x)$ for all $x \in X$. This implies that $U(-x_0) \leq T(-x_0) = -S(x_0)$. This proves the lemma. □

Theorem 2.1.1 (Hahn-Banach). *Let X be a vector space and let $S : X \to R$ be a sublinear function on X. Then there exists a linear function $L : X \to R$ such that $L(x) \leq S(x)$ for all $x \in X$.*

Proof. Define $\mathcal{F}$ to be the set

$$\{f : X \to R \text{ such that } f \text{ is sublinear and } f(x) \leq S(x) \text{ for all } x \in X\}.$$

Define the order $\prec$ on $\mathcal{F}$ by $f \prec g$ if $f(x) \geq g(x)$ for all x in X. We will prove the existence of L by proving that $\mathcal{F}$ is inductively ordered and then applying Zorn's lemma to it.

Let $\mathcal{G}$ be a totally ordered subset of $\mathcal{F}$. We will show that $\mathcal{G}$ has a maximal element. Define for w in X

$$M(w) := \inf\{g(w) : g \in \mathcal{G}\}.$$

It is clear that $g \prec M$ for any $g \in \mathcal{G}$. Also note that for any element f in $\mathcal{F}$ and for any element w in X, $0 = f(w - w) \le f(w) + f(-w)$. Therefore, $f(w) \ge -f(-w) \ge -S(-w) > -\infty$. Therefore, for any element $w \in X$, $M(w) > -\infty$. As $M(w) \le S(w) < \infty$ for all $w \in X$ it follows that M is real valued.

Let α be a non-negative real number. Then for any $w \in X$ we have

$$\begin{aligned} M(\alpha w) &= \inf\{g(\alpha w) : g \in \mathcal{G}\} \\ &= \inf\{\alpha g(w) : g \in \mathcal{G}\} = \alpha \inf\{g(w) : g \in \mathcal{G}\} \\ &= \alpha M(w). \end{aligned}$$

The second equality follows from the sublinearity of elements in $\mathcal{G}$. We now show that M is subadditive. Let h and g be any two elements of $\mathcal{G}$. Then as $\mathcal{G}$ is totally ordered either $g \prec h$ or $h \prec g$. Let x and y be any two elements in X. If $h \prec g$ then $h(x) + g(y) \ge g(x) + g(y) \ge g(x + y)$, and if $g \prec h$ then $h(x) + g(y) \ge h(x) + h(y) \ge h(x + y)$. In either case there exists a function $\overline{h}$ in $\mathcal{G}$ such that $h(x) + g(y) \ge \overline{h}(x + y)$. It follows that

$$\begin{aligned} M(x) + M(y) &= \inf\{h(x) + g(y) : h, g \in \mathcal{G}\} \\ &\ge \inf\{\overline{h}(x + y) : \overline{h} \in \mathcal{G}\} \\ &= M(x + y). \end{aligned}$$

Thus M is a real valued sublinear function. It is clear that $M(x) \le S(x)$ for all $x \in X$ and that $g \prec M$ for any $g \in \mathcal{G}$. Therefore, we have shown that every totally ordered subset of $\mathcal{F}$ has a maximal element which implies that $\mathcal{F}$ is an inductively ordered set.

Applying Zorn's lemma (Lemma 1.1.1) to $\mathcal{F}$ we have that there exists an element L in $\mathcal{F}$ such that if $L \prec f$ then $L = f$. As $L \in \mathcal{F}$ it follows that L is sublinear and $S \prec L$.

Let x_0 be any element of X. Then from Lemma 2.1.2 we have that there exists a real valued sublinear function U such that for all $y \in X$, $U(y) \le L(y) \le S(y)$ and $U(-x_0) \le -L(x_0)$. As L is the maximal element in $\mathcal{F}$, and $L \prec U$ it follows that $U = L$. This implies that $L(-x_0) = -L(x_0)$. As x_0 was chosen arbitrarily it follows that $L(-x) = -L(x)$ for all $x \in X$.

If α is a negative real number then $-\alpha L(x) = L(-\alpha x) = -L(\alpha x)$ (the first equality is true because L is sublinear and $-\alpha > 0$). Therefore, $L(\alpha x) = \alpha L(x)$ for any $\alpha \in R$ and for any $x \in X$.

Suppose, x_1 and x_2 are elements of X. Then

$$\begin{aligned} L(-x_1 - x_2) &= -L(x_1 + x_2) \\ &\ge -L(x_1) - L(x_2) \\ &= L(-x_1) + L(-x_2) \ge L(-x_1 - x_2). \end{aligned}$$

It follows that for elements x_1 and x_2 in X, $L(-x_1 - x_2) = L(-x_1) + L(-x_2)$ which implies that if x_1 and x_2 are any two elements in X then $L(x_1 + x_2) = L(x_1) + L(x_2)$. Thus we have shown that L is a linear function such that $L(x) \le S(x)$ for all x in X. □

Lemma 2.1.3. *Let X be a vector space with a sublinear function $S : X \to R$. Let x_0 be any element in X. Then there exists a linear function $L : X \to R$ such that $L(x) \le S(x)$ for all $x \in X$ and $L(x_0) = S(x_0)$.*

Proof. For all w in X let

$$U(w) := \inf\{S(w + \alpha x_0) - \alpha S(x_0) : \alpha \ge 0\}.$$

From Lemma 2.1.2 we have that U is a real valued sublinear function such that $U(x) \le S(x)$ for all $x \in X$ and $U(-x_0) \le -S(x_0)$. From Theorem 2.1.1 it follows that there exists a linear function $L : X \to R$ such that $L(x) \le U(x) \le S(x)$ for all x in X and $L(-x_0) \le U(-x_0) \le -S(x_0)$. This implies that $L(x_0) \ge S(x_0)$. Therefore, $L(x_0) = S(x_0)$ which proves the lemma. □

Theorem 2.1.2. *Let X be a vector space with $S : X \to R$ a sublinear function defined on X. Let Y be a subspace of X. Let $M : Y \to R$ be a linear function on Y such that for all $y \in Y$, $M(y) \le S(y)$. For any w in X let*

$$U(w) := \inf\{S(w - y) + M(y) : y \in Y\}.$$

Then U is a real valued sublinear function on X such that $U(x) \le S(x)$ for all x in X and $U(y) \le M(y)$ for all $y \in Y$. Also, there exists a linear function $L : X \to R$ such that $L(x) \le S(x)$ for all x in X and $L(y) = M(y)$ for all y in Y.

Proof. For any v in X let

$$\begin{aligned} T(v) &:= M(v) \text{ if } v \in Y, \\ &:= \infty \quad \text{ if } v \in X \backslash Y. \end{aligned}$$

It can be shown that if x_1 and x_2 are any two elements in X then $T(x_1+x_2) \le T(x_1)+T(x_2)$ and if $\alpha > 0$ is a real number then $T(\alpha x) = \alpha T(x)$. Note that for any w in X

$$U(w) = \inf\{S(w - y) + M(y) : y \in Y\} = \inf\{S(w - v) + T(v) : v \in X\},$$

because $T(v) = \infty$ if v is not in Y. Also, $U(0) = \inf\{S(-y) + M(y) : y \in Y\} \ge \inf\{M(-y) + M(y) : y \in Y\} = 0 > -\infty$. Thus all conditions on T, S and U stipulated in Lemma 2.1.1 are satisfied and therefore U is a real-valued sublinear function such that for all x in X, $U(x) \le S(x)$ and $U(x) \le T(x)$.

Applying Theorem 2.1.1 to U we deduce that there exists a linear function $L : X \to R$ such that for all x in X, $L(x) \le U(x)$. In particular, for all y in Y, $L(y) \le U(y) \le T(y) = M(y)$. As Y is a vector space we know that for all y in Y, $L(y) \le M(y)$ and $L(-y) \le M(-y)$. As M and L are linear we have that for all y in Y, $L(y) = M(y)$. This proves the lemma. □

2.2 Dual Spaces

In this section we study the set of bounded linear functions on a normed vector space. As will be seen the initial topology induced by the functions in this set leads to a desirable topology on vector spaces.

Theorem 2.2.1 (Bounded linear operators). *The set of all bounded linear operators between normed vector spaces* $(X, ||\cdot||_X)$ *and* $(Y, ||\cdot||_Y)$ *is denoted by* $B(X,Y)$. *Thus* $B(X,Y)$ *is the set*

$$\{T : T \text{ is linear and there exists } K \in R \text{ such that } ||Tx||_Y \leq K||x||_X\}.$$

Associate to each $T \in B(X,Y)$ *the number*

$$||T|| := \sup\{||Tx||_Y : ||x||_X \leq 1\}. \tag{2.1}$$

This definition of $||T||$ *makes* $B(X,Y)$ *into a normed vector space. If* Y *is a Banach space then so is* $B(X,Y)$ *with the norm defined in (2.1).*

Proof. The proof of the assertion that $B(X,Y)$ is a normed linear space with the norm defined by (2.1) is left as an exercise.

Suppose, $\{T_n\}$ is a Cauchy sequence in $B(X,Y)$. Then given $\epsilon > 0$ there exists an integer N such that $n, m > N$ implies that $||T_n - T_m|| \leq \epsilon$. This implies that $\sup\{||T_n(x) - T_m(x)||_Y : ||x||_X \leq 1\} \leq \epsilon$. It follows that for a given $\epsilon > 0$ and for all $x \in X$ there exists an integer N such that $n, m > N$ implies that $||T_n(x) - T_m(x)||_Y \leq \epsilon ||x||_X$. Therefore for all $x \in X$, $\{T_n(x)\}$ is a Cauchy sequence in Y. Thus, for every element $x \in X$ there exists an element $T(x)$ in Y to which $T_n(x)$ converges. Now we show that $T : X \to Y$ is a linear map.

Note that, for any integer n, $||T(x_1) + T(x_2) - T(x_1 + x_2)||_Y \leq ||T(x_1) - T_n(x_1)||_Y + ||T(x_2) - T_n(x_2)||_Y + ||T_n(x_1) + T_n(x_2) - T(x_1 + x_2)||_Y$. From the linearity of T_n and continuity of $||.||_X$, we know that the right hand side of the above inequality goes to zero as $n \to \infty$. Therefore, $T(x_1) + T(x_2) = T(x_1 + x_2)$. Also, $||\alpha T_n(x) - T(\alpha x)||_Y = ||T_n(\alpha x) - T(\alpha x)||_Y$. From the definition of $T(\alpha x)$ the right hand side goes to zero as $n \to \infty$ and therefore $\alpha T(x) = T(\alpha x)$. This proves that T is linear. Now, we show that T is bounded.

As $\{T_n\}$ is a Cauchy sequence there exists an integer M such that $m, n \geq M$ implies that $||T_m - T_n|| \leq 1$. Therefore if $m \geq M$ then $||T_m|| \leq 1 + ||T_M||$ which implies that if $m \geq M$ then $||T_m(x)||_Y \leq (1 + ||T_M||)||x||_X$ for all $x \in X$. Given any $x \in X$ and $\epsilon > 0$, choose m such that $m \geq M$ and $||T_m(x) - T(x)||_Y \leq \epsilon$. Therefore, $||T(x)||_Y \leq (1 + ||T_M||)||x||_X + \epsilon$. As ϵ is arbitrary it follows that $||T(x)||_Y \leq (1 + ||T_M||)||x||_X$. This relation holds for any $x \in X$. Therefore T is bounded.

Given any real $\epsilon > 0$, choose M_1 such that $n, m \geq M_1$ implies that $||T_n - T_m|| \leq \frac{\epsilon}{2}$. This implies that for all $x \in X$, if $n, m \geq M_1$ then $||T_n(x) - T_m(x)||_Y \leq \frac{\epsilon}{2}||x||_X$. Given any x choose $n \geq M_1$, large enough so that

$||T_n(x) - T(x)||_Y \leq \frac{\epsilon}{2}||x||_X$. Therefore, if $m \geq M_1$ then $||T(x) - T_m(x)||_Y \leq \epsilon||x||_X$. This implies that $||T - T_m|| \leq \epsilon$ if $m > M_1$. As ϵ is arbitrary it follows that $||T - T_m|| \to 0$. This proves that T_n converges to T in the norm topology. Thus we have established that any Cauchy sequence in $B(X, Y)$ converges if Y is Banach. □

If X is not the empty set or $\{0\}$ then it is also true that $||T|| = \sup\{||Tx||_Y : ||x||_X = 1\}$ and $||T|| = \sup\{\frac{||Tx||_Y}{||x||_X} : x \neq 0\}$. Proof of this assertion is left to the reader.

Corollary 2.2.1 (Hahn-Banach: extension form). *Let Y be a subspace of a normed vector space $(X, ||\cdot||_X)$. If $f : Y \to R$ be a bounded linear function on Y then there exists a linear function $F : X \to R$ such that $F(y) = f(y)$ for all $y \in Y$ and $||F|| = ||f||$.*

Proof. Let $S : X \to R$ be defined by $S(x) = ||f|| \, ||x||_X$. Then it is clear that S is a real valued sublinear function. It is also true that for all $y \in Y$, $|f(y)| \leq S(y)$ (because $||f|| = \sup\{|f(y)| : y \in Y, ||y||_X \leq 1\}$). From Theorem 2.1.2 it follows that there exists a linear function $F : X \to R$ such that

$$F(x) \leq S(x) \text{ for all } x \in X, \tag{2.2}$$

and

$$F(y) = f(y) \text{ for all } y \in Y. \tag{2.3}$$

Equation (2.2) implies that $||F|| \leq ||f||$ and equation (2.3) implies that $||F|| \geq ||f||$. Therefore, $||F|| = ||f||$. □

Definition 2.2.1 (Dual spaces). *Let $(X, ||\cdot||_X)$ be a normed vector space. Then X^* denotes the set of bounded linear functions from X to R. The norm on any element x^* in X^* is defined as*

$$||x^*|| := \sup\{x^*(x) : ||x||_X \leq 1\}. \tag{2.4}$$

In other words we have $X^* = B(X, R)$. X^* is said to be the dual space of X.

It follows immediately from Theorem 2.2.1 that the dual space of any normed vector space is Banach (because R is Banach).

Theorem 2.2.2. *Let $x_0 \in X$ where $(X, ||\cdot||_X)$ is a normed vector space with X^* as its dual. Then there exists an element $x_0^* \in X^*$ such that $||x_0^*|| = 1$ and $< x_0, x_0^* >= ||x_0||_X$.*

Proof. Let $S : X \to R$ be defined by $S(x) := ||x||_X$ for all $x \in X$. Then S is a real valued sublinear function. From Lemma 2.1.3 we know that there exists a linear function $L : X \to R$ such that $L(x) \leq S(x)$ for all $x \in X$ and $L(x_0) = S(x_0)$. This implies that $||L|| = 1$ and $L(x_0) = ||x_0||_X$. Denote L by x_0^*. This proves the theorem. □

Theorem 2.2.3. *Let $(X, ||\cdot||_X)$ and $(Y, ||\cdot||_Y)$ be normed vector spaces. Let $X \times Y$ be endowed with the product norm. Then there exists an isometric isomorphism between $(X \times Y)^*$ and $X^* \times Y^*$ with the norm on $X^* \times Y^*$ defined by*

$$||(x^*, y^*)|| := \max\{||x^*||, ||y^*||\},$$

where $x^* \in X^*$ and $y^* \in Y^*$.

Proof. Let f be a bounded linear real valued function on $X \times Y$. Then from the linearity of f we have for any $(x, y) \in X \times Y$,

$$f((x,y)) = f((x,0) + (0,y)) = f((x,0)) + f((0,y)). \tag{2.5}$$

It is clear that x^* a function on X defined by $x^*(x) = f((x,0))$ is linear. Similarly, y^* a function on Y defined by $y^*(y) = f((0,y))$ is linear. Also, from equation (2.5) we have $|x^*(x) + y^*(y)| = |f(x,y)| \leq ||f||(||x||_X + ||y||_Y)$ for all $(x,y) \in X \times Y$. In particular, $|x^*(x)| \leq ||f||\,||x||_X$ which implies that $||x^*|| \leq ||f||$. Similarly, it can be shown that $||y^*|| \leq ||f||$ and therefore $\max\{||x^*||, ||y^*||\} \leq ||f||$.

Given any $\epsilon > 0$ we know that there exists $x_\epsilon \in X$ and $y_\epsilon \in Y$ such that $||x_\epsilon||_X + ||y_\epsilon||_Y \leq 1$ and $|f(x,y)| \geq ||f|| - \epsilon$. Therefore, for every $\epsilon > 0$ there exists $x_\epsilon \in X$ and $y_\epsilon \in Y$ such that $||x_\epsilon||_X + ||y_\epsilon||_Y \leq 1$ and $|x^*(x) + y^*(x)| \geq ||f|| - \epsilon$ which implies that $||x^*||\,||x_\epsilon||_X + ||y^*||\,||y_\epsilon||_Y \geq ||f|| - \epsilon$. Thus, $\max\{||x^*||, ||y^*||\}(||x_\epsilon||_X + ||y_\epsilon||_Y|) \geq ||f|| - \epsilon$. As $||x_\epsilon||_X + ||y_\epsilon||_Y \leq 1$ and $\epsilon > 0$ is arbitrary it follows that $\max\{||x^*||, ||y^*||\} \geq ||f||$.

Thus we have established that the map $i : (X \times Y)^* \to X^* \times Y^*$ which takes $f \in (X \times Y)^*$ into $(x^*, y^*) \in X^* \times Y^*$ as defined above is isometric. The fact that i is isomorphic is left to the reader to prove. □

Definition 2.2.2 (Second dual space). *Let $(X, ||\cdot||_X)$ be a normed vector space with $(X^*, ||\cdot||)$ as its dual. The set of all bounded linear functions from X^* to R is the second dual of X. It is denoted as X^{**}.*

Every element x in X can be thought of as a map from X^* into R. Thus every element $x \in X$ can be identified with an element in X^{**}. For any x^* in X^* let $(J(x))(x^*) := x^*(x)$. Note that $(J(x))(x_1^* + x_2^*) = (x_1^* + x_2^*)(x) = x_1^*(x) + x_2^*(x) = (J(x))(x_1^*) + (J(x))(x_2^*)$ where $x \in X$, $x_1^* \in X^*$ and $x_2^* \in X^*$. Also, $(J(x))(\alpha x^*) = (\alpha x^*)(x) = \alpha x^*(x) = \alpha(J(x))(x^*)$. Therefore, it follows that $J(x)$ is a linear map from X^* to R. It is also a bounded linear function on X^*. Indeed, for $x \in X$ and $x^* \in X^*$, $|(J(x))(x^*)| = |x^*(x)| \leq ||x^*||\,||x||_X$. Therefore,

$$\sup\{(J(x))(x^*) : ||x^*|| \leq 1\} \leq ||x||_X. \tag{2.6}$$

Thus we have established that

$$||J(x)|| \leq ||x||_X. \tag{2.7}$$

In fact we will show using a Hahn-Banach result that $||J(x)|| = ||x||_X$. We call the map $J : X \to X^{**}$ the *canonical map* from X to X^{**}. Let $(X, ||\cdot||_X)$ be a normed vector space with X^* as its dual. We use the symmetric notation $< x, x^* >$ to denote $x^*(x)$. With this notation we have $(J(x))(x^*) =< x^*, J(x) >=< x, x^* >$. We will call $< .,. >$ the *bilinear form* on X.

Definition 2.2.3 (Adjoint map). *Let $A : (X, ||\cdot||_X) \to (Y, ||\cdot||_Y)$ be a bounded linear map from a normed vector space X to a normed vector space Y. The adjoint of the map A, denoted by A^* is a map from Y^* to X^* defined by*

$$< x, A^*(y^*) >_X := < A(x), y^* >_Y \quad \textit{for all } x \in X, \textit{ and for all } y^* \in Y^*,$$

where $< .,. >_X$ is the bilinear form on X and $< .,. >_Y$ is the bilinear form on Y.

Theorem 2.2.4. *For any element x_0 of a normed linear space $(X, ||\cdot||_X)$ there exists an element x_0^* in X^* such that $||x_0^*|| \leq 1$ and $| < x_0, x_0^* > | = ||x_0||_X$. Also, if $J : X \to X^{**}$ is the canonical map then $||J(x)|| = ||x||_X$.*

Proof. Note that $||\cdot||_X : X \to R$ is a sublinear function on X. Applying Lemma 2.1.3 we know that there exists a linear function $L : X \to R$ such that

$$L(x) \leq ||x||_X \text{ for all } x \in X \text{ and } L(x_0) = ||x_0||_X.$$

Because L is linear it follows that $|L(x)| \leq ||x||_X$ for all x in X which implies that $||L|| \leq 1$. Therefore, L is in X^* and we denote L by x_0^*. Thus $||x_0^*|| \leq 1$ and $< x_0, x_0^* >= ||x_0||_X$.

We have established earlier that $||J(x_0)|| \leq ||x_0||_X$. However, $||J(x_0)|| = \sup\{< x^*, J(x_0) >: ||x^*|| \leq 1\} \geq < x_0^*, J(x_0) >=< x_0, x_0^* >= ||x_0||_X$. This proves that $||J(x)||_X = ||x||_X$ for all x in X. □

Theorem 2.2.5. *Let $(X, ||\cdot||_X)$ and $(Y, ||\cdot||_Y)$ be normed vector spaces with a linear map A defined from X to Y. Let $A^* : Y^* \to X^*$ denote the adjoint of A. Then, $||A|| = ||A^*||$ that is $\sup\{||A(x)||_Y : x \in X \text{ and } ||x||_X \leq 1\} = \sup\{||A^*(y^*)|| : y^* \in Y^* \text{ and } ||y^*|| \leq 1\}$.*

Proof. The proof is similar to the proof of Theorem 2.2.4 and is left to the reader. □

Definition 2.2.4 (Reflexive). *Let $(X, ||\cdot||_X)$ be a normed vector space. We say X is reflexive if $J(X) = X^{**}$. Thus for a reflexive space every element x^{**} in X^{**} can be identified with an element x in X.*

2.3 Weak Topologies

We study here the weakest topology on X which makes all bounded linear function on X continuous and the weakest topology on X^* which makes all the elements of X viewed as functions on X^* continuous.

Definition 2.3.1 (Weak topology, Weak-star topology). *Let $(X, ||\cdot||_X)$ be a normed linear space with X^* as its dual. The initial topology on X induced by the elements of X^* is called the weak topology on X and is denoted by $W(X, X^*)$.*

The initial topoogy on X^* induced by the set

$$J(X) := \{J(x) : x \in X\},$$

where $J : X \to X^{**}$ is the canonical map is called the weak-star topology on X^* and is denoted by $W(X^*, X)$.

Let A be any open set in R. As A is open, for every $r \in A$ there exists a positive number $\epsilon_r > 0$ such that the set $\{t : |t - r| < \epsilon_r\}$ is a subset of A. Therefore,

$$A = \bigcup_{r \in A} \{t : |t - r| < \epsilon_r\}.$$

If x_i^* is an element of X^* then $x_i^* : X \to R$ and a subbase for the weak topology for X (see Lemma 1.2.6) are sets of the form $\{x \in X : | < x, x^* > -r| < \epsilon\}$ where $x^* \in X^*$, $r \in R$ and $\epsilon > 0$. This implies that if N is a neighbourhood of x in the weak topology then it must contain a set of the form

$$\bigcap_{i=1}^{n} \{y \in X : | < y - x, x_i^* > | < \epsilon_i\},$$

where n is an integer, $x_1^*, \ldots, x_n^*$ are elements of X^* and $\epsilon_1, \ldots, \epsilon_n$ are positive real numbers.

Similarly if N^* is a neighbourhood of x^* in the weak-star topology then it must contain a set of the form

$$\bigcap_{i=1}^{n} \{y^* \in X^* : | < x_i, y^* - x^* > | < \epsilon_i\},$$

where n is an integer, $x_1, \ldots, x_n$ are elements of X and $\epsilon_1, \ldots, \epsilon_n$ are positive real numbers.

Lemma 2.3.1. *Let $\{x_\lambda\}$ be a net in X, then $x_\lambda \to x_0$ in the weak topology if and only if*

$$< x_\lambda, x^* > \to < x_0, x^* > \quad \text{in } R \text{ for all } x^* \text{ in } X^*.$$

Similarly, if $\{x_\lambda^*\}$ is a net in X^*, then $x_\lambda^* \to x_0^*$ in the weak-star topology if and only if

$$< x, x_\lambda^* > \to < x, x_\lambda^* > \quad \text{in } R \text{ for all } x \text{ in } X.$$

Proof. Follows from Lemma 1.2.7. □

Theorem 2.3.1 (Banach-Alaoglu). *Let* $(X, ||\cdot||_X)$ *be a normed vector space with* X^* *as its dual. The set*

$$B^* := \{x^* : ||x^*|| \leq M\}, \tag{2.8}$$

is compact in the weak-star topology for any $M \in R$.

Proof. Let x^*_λ be a universal net in B^* on the directed set Λ. For any x in X, $(J(x))(x^*_\lambda) =< x, x^*_\lambda >$ is a universal net in R (see Lemma 1.2.11) and furthermore, $| < x, x^*_\lambda > | \leq M||x||_X$. Therefore $< x, x^*_\lambda >$ is a universal net in the compact set $B := \{r \in R : -M||x||_X|| \leq r \leq M||x||_X\}$. Therefore, there exists an element $f(x)$ in R such that $< x, x^*_\lambda > \rightarrow f(x)$ and $f(x) \in B$ (see Theorem 1.2.3).

Let x and y be any two elements in X. Let $< x, x^*_\lambda > \rightarrow f(x)$ and $< y, x^*_\lambda > \rightarrow f(y)$. Also, it is true that $< x + y, x^*_\lambda > \rightarrow f(x + y)$. This implies that $< x, x^*_\lambda > + < y, x^*_\lambda > \rightarrow f(x + y)$. Therefore, $f(x + y) = f(x) + f(y)$. It can be shown in a similar manner that $f(\alpha x) = \alpha f(x)$ for any real number α and x in X. This implies $f : X \rightarrow R$ is linear. As $f(x)$ is in B it follows that $||f|| := \sup\{|f(x)| : ||x||_X \leq 1\} \leq M$. Therefore, f belongs to B^* and by definition $< x, x^*_\lambda > \rightarrow f(x)$ for all x in X. Hence, from Lemma 2.3.1 it follows that $x^*_\lambda \rightarrow f$. Thus, every universal net in B^* is convergent in the weak-star topology which implies that B^* is compact in the weak-star topology (see Theorem1.2.3). □

This shows that the weak-star topology is important because, unlike the norm topology, the unit norm ball is compact. However, it is not true that every sequence in a compact topological space has a convergent subsequence (an example is given later). The following result guarantees existence of a convergent subsequence when the space is separable.

Theorem 2.3.2. *Let* $(X, ||\cdot||_X)$ *be a separable normed vector space with* X^* *as its dual. Then every sequence in* $\{x^* : ||x^*|| \leq M\}$ *has a convergent subsequence in the weak-star topology where* $M \in R$.

Proof. Left to the reader. □

Next we present a result on the compactness of the norm ball of X in the weak topology.

Theorem 2.3.3. *Let* $(X, ||\cdot||_X)$ *be a normed vector space with* X^* *as its dual. The set*

$$B := \{x : ||x||_X \leq 1\}, \tag{2.9}$$

is compact in the weak topology if and only if X *is reflexive.*

2.4 ℓ_p Spaces

In this section we study the vector space of sequences with different norms imposed on it. We denote the space of sequences by ℓ. Therefore, every element of ℓ is a function from the set of integers to the real numbers; $\ell = \{x | x : I \to R\}$. If $\alpha \in R$ and $x \in \ell$ then we define αx by $(\alpha x)(i) = \alpha x(i)$. For $x \in \ell$ and $y \in \ell$ we define $x + y$ by $(x + y)(i) = x(i) + y(i)$. It is evident that ℓ with the scalar multiplication and vector addition as defined above is a vector space. Often, for an element $x \in \ell$ and $i \in I$, $x(i)$ is denoted as x_i. For an element x in ℓ define

$$||x||_p := \left(\sum_{i=-\infty}^{\infty} |x_i|^p \right)^{1/p} \quad \text{and} \quad ||x||_\infty := \sup_{-\infty < i < \infty} |x_i|,$$

where $0 < p < \infty$ is a real number.

Definition 2.4.1 (ℓ_p spaces). *Let p be a real number such that $0 < p \leq \infty$. The ℓ_p spaces are defined by*

$$\ell_p := \{x \in \ell : ||x||_p < \infty\}.$$

We will restrict our attention to right-sided sequences i.e. for elements of ℓ which map negative integers to zero. This is a matter of convenience and the results in this section are valid in general.

Definition 2.4.2. *c_0 is a subset of ℓ such that for every $x \in c_0$,* $\lim x_i = 0$.

Now we show that ℓ_p spaces are nested.

Lemma 2.4.1. *If $0 < p < q \leq \infty$ then $\ell_p \subset c_0$ and $\ell_p \subset \ell_q$.*

Proof. We leave the first part of the proof to the reader. Suppose $x \in \ell_p$. Then $\sum_{i=0}^{\infty} |x_i|^p < \infty$. Therefore, there exists an integer N such that $i > N$ implies that $|x_i| < 1$. Note that if $p \leq q$ then for $i > N$, $|x_i|^q \leq |x_i|^p$. Therefore,

$$\begin{aligned} \sum_{i=0}^{\infty} |x_i|^q &= \sum_{i=0}^{N} |x_i|^q + \sum_{i=N+1}^{\infty} |x_i|^q \\ &\leq \sum_{i=0}^{N} |x_i|^q + \sum_{i=N+1}^{\infty} |x_i|^p \\ &\leq \sum_{i=0}^{N} |x_i|^q + ||x||_p^p < \infty \end{aligned}$$

Thus $x \in \ell_q$. □

Lemma 2.4.2 (Holder's inequality). *Suppose that* $1 \le p \le \infty$ *and* $\frac{1}{p} + \frac{1}{q} = 1$, *then for elements* x *and* y *in* ℓ

$$||xy||_1 \le ||x||_p ||y||_q, \tag{2.10}$$

where $xy \in \ell$ *is defined by* $xy(i) = x(i)y(i)$ *with* i *being any integer. Also, the equality holds in* (2.10) *if and only if* $(\frac{|x_i|}{||x_p||})^p = (\frac{|y_i|}{||y_q||})^q$ *for all* $i \ge 0$.

Proof. We will assume that $1 < p < \infty$ and $1 < q < \infty$. The other cases are left to the reader to prove. We will first show that if $r \ge 0$, $t \ge 0$ and $0 < \lambda < 1$ then

$$r^\lambda t^{(1-\lambda)} \le \lambda r + (1-\lambda)t. \tag{2.11}$$

Consider the function f

$$f(h) = h^\lambda - \lambda h + \lambda - 1,$$

which maps the set of positive real numbers to R. Then $f'(h) = \lambda(h^{(\lambda-1)} - 1)$. Therefore for $0 < h < 1$, $f'(h) > 0$ and for $h > 1$, $f'(h) < 0$. This implies that f is a strictly increasing function in the region $0 < h < 1$ and therefore $f(h) < f(1) = 0$ if $0 < h < 1$. Similarly in the region $h > 1$, f is an strictly decreasing function and $f(h) < f(1) = 0$ if $h > 1$. This implies that $f(h) \le 0$ if $h \ge 0$ and the equality holds only if $h = 1$. In other words

$$h^\lambda \le \lambda h + 1 - \lambda,$$

with equality only if $h = 1$. If $t > 0$ then by substituting $\frac{r}{t}$ for h we have the inequality in (2.11). If $t = 0$ then the inequality in (2.11) clearly holds.

Let

$$r = \left(\frac{|x_i|}{||x||_p}\right)^p, t = \left(\frac{|y_i|}{||y||_q}\right)^q, \lambda = \frac{1}{p} \text{ and } 1 - \lambda = \frac{1}{q}.$$

From inequality (2.11) we have

$$\frac{|x_i|}{||x||_p}\frac{|y_i|}{||y||_q} \le \frac{1}{p}\left(\frac{|x_i|}{||x||_p}\right)^p + \frac{1}{q}\left(\frac{|y_i|}{||y||_q}\right)^q. \tag{2.12}$$

Therefore,

$$\sum_{i=0}^{\infty} \frac{|x_i|}{||x||_p}\frac{|y_i|}{||y||_q} \le \frac{1}{p}\sum_{i=0}^{\infty}\left(\frac{|x_i|}{||x||_p}\right)^p + \frac{1}{q}\sum_{i=0}^{\infty}\left(\frac{|y_i|}{||y||_q}\right)^q = \frac{1}{p} + \frac{1}{q} = 1.$$

This proves the lemma. □

Lemma 2.4.3 (Minkowski's inequality). *Suppose that* $1 \le p \le \infty$ *then for elements* x *and* y *in* ℓ

$$||x + y||_p \le ||x||_p + ||y||_p.$$

The equality holds if and only if there exists a real k *such that* $x_i = ky_i$ *for all* $i = 0, \dots n$.

Proof. We will assume that $1 < p < \infty$ and leave the other cases for the reader to prove. Let q be such that $\frac{1}{p} + \frac{1}{q} = 1$. Note that

$$\begin{aligned}\sum_{i=0}^{N} |x_i + y_i|^p &= \sum_{i=0}^{N} |x_i + y_i|^{p-1}|x_i| + \sum_{i=0}^{N} |x_i + y_i|^{p-1}|y_i| \\ &\leq \left(\sum_{i=0}^{N} |x_i + y_i|^{(p-1)q}\right)^{\frac{1}{q}} \left[\left(\sum_{i=0}^{N} |x_i|^p\right)^{\frac{1}{p}} + \left(\sum_{i=0}^{N} |y_i|^p\right)^{\frac{1}{p}}\right] \\ &= \left(\sum_{i=0}^{N} |x_i + y_i|^p\right)^{\frac{1}{q}} \left[\left(\sum_{i=0}^{N} |x_i|^p\right)^{\frac{1}{p}} + \left(\sum_{i=0}^{N} |y_i|^p\right)^{\frac{1}{p}}\right],\end{aligned}$$

where the first inequality follows by applying Holder's inequality (see Lemma 2.4.2), whereas the second equality is true because $(p-1)q = p$. Also, note that the equality holds if and only if (see Lemma 2.4.2),

$$\frac{|x_i + y_i|^{(p-1)q}}{\sum_{i=0}^{N} |x_i + y_i|^{(p-1)q}} = \frac{|x_i|^p}{\sum_{i=0}^{N} |x_i|^p} = \frac{|y_i|^p}{\sum_{i=0}^{N} |y_i|^p}.$$

Therefore the equality holds if and only if there exists a real k such that $x_i = ky_i$ for all $i = 0, \ldots n$.

Dividing the above inequality by $\left(\sum_{i=0}^{N} |x_i + y_i|^p\right)^{\frac{1}{q}}$ and noting that $\frac{1}{p} + \frac{1}{q} = 1$, we have

$$\left(\sum_{i=0}^{N} |x_i + y_i|^p\right)^{\frac{1}{p}} \leq \left(\sum_{i=0}^{N} |x_i|^p\right)^{\frac{1}{p}} + \left(\sum_{i=0}^{N} |y_i|^p\right)^{\frac{1}{p}}.$$

This holds for all positive integers N. Therefore, $||x+y||_p \leq ||x||_p + ||y||_p$. □
It follows from Lemma 2.4.3 that $||\cdot||_p$ is a norm on ℓ_p for $1 \leq p \leq \infty$.

Theorem 2.4.1. *The following statements are true.*

1. *If* $1 \leq p \leq \infty$ *then* $(\ell_p, ||\cdot||_p)$ *is a Banach space.*
2. *If* $0 < p < 1$ *then* $(\ell_p, ||\cdot||_p)$ *is complete.*

Proof. (1) Let $\{x^n\}$ be a Cauchy sequence in ℓ_p with $1 \leq p < \infty$. Then given any $\epsilon > 0$ there exists an M such that $n, m > N$ implies that

$$|x^n(i) - x^m(i)|^p \leq ||x_n - x_m||^p \leq \epsilon.$$

Therefore, $\{x^n(i)\}$ is a Cauchy sequence in R. This implies that there exists an element $x(i)$ such that $x^n(i) \to x(i)$ because R is complete. We will show now that $x := \{x(0), x(1), x(2), \ldots, \}$ is in ℓ_p. Indeed, from Lemma 1.4.2 we know that there exists an $M \in R$ such that for all $n = 1, \ldots, \infty$

$$||x^n||_p^p \le M.$$

This implies that for all $n = 1, \ldots, \infty$

$$\sum_{i=0}^{K} |x^n(i)|^p \le M,$$

where K is any positive integer. Taking limits as $n \to \infty$ it follows that

$$\sum_{i=0}^{K} |x(i)|^p \le M,$$

for any positive integer K. Therefore, $x \in \ell_p$. We prove now that x^n converges to x in $||\cdot||_p$ norm. Given any $\epsilon > 0$ there exists N such that $n, m > N$ implies that $||x^n - x^m||_p^p \le \epsilon$. This implies that for any integer K and $n, m \ge N$

$$\sum_{i=0}^{K} |x^n(i) - x^m(i)|^p \le \epsilon.$$

Letting $m \to \infty$ we have that for any integer K and $n \ge N$

$$\sum_{i=0}^{K} |x^n(i) - x(i)|^p \le \epsilon.$$

This proves that $||x^n - x||_p \to 0$ as $n \to \infty$. Thus we have established that if $1 \le p \le \infty$ then $(\ell_p, ||\cdot||_p)$ is a Banach space.

If $p = \infty$ and $\{x^n\}$ is a Cauchy sequence in ℓ_∞, then given any ϵ there exists a N such that for $i = 1, \ldots, \infty$ and for any $n, m \ge N$

$$|x^n(i) - x^m(i)| \le \epsilon. \tag{2.13}$$

This implies that $\{x^n(i)\}$ is a Cauchy sequence in R and we suppose it converges to $x(i)$. Letting $m \to \infty$ in (2.13) we see that given any $\epsilon > 0$ there exists a N such that if $n > N$ then

$$|x^n(i) - x(i)| \le \epsilon,$$

for any $i = 1, \ldots, \infty$. This proves that $x^n \to x$ in the $||\cdot||_\infty$ norm. Thus we have established that if $1 \le p \le \infty$ then $(\ell_p, ||\cdot||_p)$ is a Banach space.

(2) Proof is identical to the proof for $1 \le p \le \infty$. ℓ_p is not a Banach space for $0 < p < 1$ because the Minkowski's inequality does not hold. □

Lemma 2.4.4. *If $0 < p < \infty$ then $(\ell_p, ||\cdot||_p)$ is separable. Also, $(c_0, ||\cdot||_\infty)$ is separable.*

Proof. Let A be a subset of ℓ such that for every element $x \in A$, there exists an integer N such that if $i \ge N$ then $x(i) = 0$ and $x(i)$ is rational for all $i = 0, \ldots, \infty$. We will show that A is dense in $(\ell_p, ||\cdot||_p)$. Let x be any element in ℓ_p. Given any $\epsilon > 0$ there exists an integer M such that

$$\sum_{i=M+1}^{\infty} |x(i)|^p \le \frac{\epsilon}{2}.$$

For every $x(i)$, with $i \le M$ there exists a rational number $r(i)$ such that $|x(i) - r(i)|^p \le \frac{\epsilon}{2^{i+2}}$. Define $r \in \ell_p$ as $r := \{r(0), r(1), \ldots, r(M), 0, 0, \ldots\}$. Note that $r \in A$ and

$$\begin{aligned} ||x - r||_p^p &= \sum_{i=0}^{M} |x(i) - r(i)|^p + \sum_{i=M+1}^{\infty} |x(i) - r(i)|^p \\ &\le \sum_{i=0}^{M} \frac{\epsilon}{2^{i+2}} + \sum_{i=M+1}^{\infty} |x(i)|^p \\ &\le \sum_{i=0}^{\infty} \frac{\epsilon}{2^{i+2}} + \frac{\epsilon}{2} \le \epsilon \end{aligned}$$

It can be shown that A is a countable set. Thus we have established that if $0 < p < \infty$ then ℓ_p has a countable dense subset and therefore it is separable. The proof for c_0 being separable is very similar to the one given above and is left to the reader. □

Theorem 2.4.2. *Let $1 \le p < \infty$. Then there exists an isometric isomorphism between ℓ_p^* and ℓ_q if $\frac{1}{p} + \frac{1}{q} = 1$.*

Proof. We prove the theorem for $1 < p < \infty$. The proof of the theorem when $p = 1$ is left to the reader. Let $x^* \in \ell_p^*$ be a bounded linear function on ℓ_p. Let $e^n \in \ell$ denote the sequence $\{e^n(i)\}$ such that

$$\begin{aligned} e^n(i) &= 0 \text{ if } i \ne n \\ &= 1 \text{ if } i = n. \end{aligned}$$

For any element $x = \{x(i)\}$ in ℓ_p it can be shown that $\sum_{i=0}^{N} x(i)e^i$ converges to x in the $||\cdot||_p$ norm. As x^* is continuous it follows that

$$x^*(x) = x^*(\lim_{N\to\infty} \sum_{i=0}^{N} x(i)e^i) = \lim_{N\to\infty} \sum_{i=0}^{N} x(i)x^*(e^i).$$

Let $y := \{x^*(e^0), x^*(e^1), \ldots\}$. Therefore, $x^*(x) = \lim_{N\to\infty} \sum_{i=0}^{N} x(i)y(i)$. We will show that $y \in \ell_q$. Let x^N in ℓ_p be defined by

$$\begin{aligned} x^N(i) &= |y(i)|^{\frac{q}{p}} sgn(y(i)) && \text{if } i \le N \\ &= 0 && \text{if } i > N. \end{aligned}$$

With this definition we have

$$\|x^N\|_p = \left(\sum_{i=0}^{N} |y(i)|^q\right)^{\frac{1}{p}} \tag{2.14}$$

and

$$x^*(x^N) = \sum_{i=0}^{N} x^N(i)y(i) = \sum_{i=0}^{N} |y(i)|^{\frac{q}{p}+1} = \sum_{i=0}^{N} |y(i)|^q. \tag{2.15}$$

However, $|x^*(x^N)| \leq \|x^*\| \, \|x^N\|_p$ and thus it follows from (2.14) and (2.15) that

$$\sum_{i=0}^{N} |y(i)|^q \leq \|x^*\| \left(\sum_{i=0}^{N} |y(i)|^q\right)^{\frac{1}{p}},$$

which implies that

$$\left(\sum_{i=0}^{N} |y(i)|^q\right)^{1-\frac{1}{p}} \leq \|x^*\|$$

and

$$\left(\sum_{i=0}^{N} |y(i)|^q\right)^{\frac{1}{q}} \leq \|x^*\|.$$

This holds for all N and therefore $\|y\|_q \leq \|x^*\|$. Note that $\|x^*\| := \sup\{|x^*(x)| : \|x\|_p \leq 1\}$. Therefore for any $\epsilon > 0$ there exists $x \in \ell_p$ such that $|x^*(x)| + \epsilon \geq \|x^*\|$ and $\|x\|_p \leq 1$. This implies that $\sum_{i=0}^{\infty} |x(i)| \, |y(i)| \geq \|x^*\| - \epsilon$. From Holder's inequality it follows that $\|y\|_q \|x\|_p \geq \|x^*\| - \epsilon$. As $\|x\|_p \leq 1$ and ϵ is any arbitrary positive number it follows that $\|y\|_q \geq \|x^*\|$ and therefore $\|y\|_q = \|x^*\|$. Thus we have shown that there exists a map $F : \ell_p^* \rightarrow \ell_q$ defined by

$$F(x^*) = \{x^*(e^i)\}$$

such that F is isometric and also, $x^*(x) = \sum_{i=0}^{\infty} x(i)F(x^*)(i)$. It is clear that F is one to one.

We will now show that F is onto. Indeed, let $z := \{z(i)\}$ be any element in ℓ_q. Let $f : \ell_p \rightarrow R$ be defined by $f(x) = \sum_{i=0}^{\infty} x(i)z(i)$. It is clear that f is linear and $|f(x)| = |\sum_{i=0}^{\infty} x(i)z(i)| \leq \|x\|_p \|z\|_q$. Therefore, $\|f\| \leq \|z\|_q$ and $f \in x^*$. It can be easily verified that $F(f) = z$. Thus F is a one to one and onto isometric map from ℓ_p^* to ℓ_q. □

Lemma 2.4.5. *There exists an isometric isomorphism which maps c_0^* to ℓ_1 where c_0^* is the dual space of $(c_0, ||\cdot||_\infty)$.*

Proof. Let $x^* \in c_0^*$ and let $y := \{x^*(e^i)\}$ where $e^i \in \ell$ is defined in the proof of the previous theorem. Define $x^N \in c_0$ by

$$\begin{aligned} x^N(i) &= sgn(y(i)) \text{ if } i \leq N \\ &= 0 \qquad\qquad\quad \text{if } i > N. \end{aligned}$$

Then $x^*(x^N) = \sum_{i=0}^{N} sgn(y(i))y(i) = \sum_{i=0}^{N} |y(i)|$. As $|x^*(x^N)| < ||x^*||\ ||x^N||_\infty$ we have $\sum_{i=0}^{N} |y(i)| \leq ||x^*||\ ||x^N||_\infty \leq ||x^*||$. This holds for any arbitrary N and therfore $||y||_1 \leq ||x^*||$.

The rest of the proof is similar to the one given for $1 < p < \infty$ and is left to the reader. □

Often, if there is a isometric isomorphism between two spaces X and Y the notation $X = Y$ is employed. For example $c_0^* = \ell_1$ means that there is a isometric isomorphism between the set of bounded linear functions on c_0 and ℓ_1. Thus, Theorem 2.4.2 says that $\ell_p^* = \ell_q$ where $\frac{1}{p} + \frac{1}{q} = 1$ and $1 \leq p < \infty$.

The following example illustrates that not every sequence in the set $\{x^* : ||x^*|| \leq 1\}$ has a weak-star convergent subsequence.

Example 2.4.1. Consider the normed linear space $(\ell_\infty, ||\cdot||_\infty)$ and let ℓ_∞^* denote its dual. Define $x_n^* \in \ell_\infty^*$ by

$$< x, x_n^* >= x(n).$$

It is clear that $||x_n^*|| = \sup\{| < x, x_n^* > | : ||x||_\infty \leq 1\} \leq 1$ for all integers n. Let $x_{n_k}^*$ be a subsequence of $\{x_n^*\}$. Let x be an element in ℓ_∞ such that

$$\begin{aligned} x(i) &= \ \ 1 \text{ if } i = n_k, \ k \text{ is even} \\ &= -1 \text{ if } i = n_k, \ k \text{ is odd} \\ &= \ \ 0 \text{ otherwise.} \end{aligned}$$

It is clear that $< x, x_{n_k}^* >= 1$ if k is even and $< x, x_{n_k}^* >= -1$ if k is odd. Therefore, $< x, x_{n_k}^* >$ does not converge which implies that $x_{n_k}^*$ is not a convergent sequence in ℓ_∞^*. However, it should be noted that Banach-Alaoglu theorem implies that

$$B^* := \{x^* \in \ell_\infty^*, ||x^*|| \leq 1\} \subset \ell_\infty^*$$

is weak-star compact. Therefore every universal net has a convergent subnet. But we have demonstrated that not every sequence has a convergent subsequence (however, every sequence has a convergent subnet).

3. Convex Analysis

One of the most important concepts in optimization is that of convexity. It can be said that the only true global optimization results involve convexity in one way or another. Establishing that a problem is equivalent to a finite dimensional convex optimization is often considered as solving the problem. This viewpoint is further rienforced due to efficient software packages available for convex programming.

In this chapter we study convex sets and convex functions. The study of sublinear functions in the previous chapter will aid in establishing results on convex optimization. The famous Kuhn-Tucker Lagrange duality result is established which is important because it converts a constrained optimization problem to an unconstrained optimization problem.

3.1 Convex Sets and Convex Maps

In this section we present the definitions of convex sets, convex maps and related concepts. Preliminary results on convexity are given.

Definition 3.1.1 (Convex sets). *A subset Ω of a vector space X is said to be convex if for any two elements c_1 and c_2 in Ω and for a real number λ with $0 < \lambda < 1$ the element $\lambda c_1 + (1 - \lambda)c_2 \in \Omega$ (see Figure 3.1). The set $\{\}$ is assumed to be convex.*

Lemma 3.1.1. *If C is a convex subset of a normed vector space, $(X, \|\cdot\|_X)$, then $int(C)$ is convex and $\overline{C}$ is convex where $\overline{C}$ denotes the norm closure of the set C.*

Proof. If $int(C)$ is empty then by assumption it is convex. Suppose, x_1 and x_2 are elements in $int(C)$. Then there exists an $\epsilon > 0$ such that

$$B(x_1, \epsilon) := \{x : \|x - x_1\|_X < \epsilon\}$$

and

$$B(x_2, \epsilon) := \{x : \|x - x_2\|_X < \epsilon\},$$

both are subsets of C. Let $\lambda \in R$ be such that $0 < \lambda < 1$. Let $z := \lambda x_1 + (1 - \lambda)x_2$. Then for any $w \in X$ with $\|w\|_X < \epsilon$,

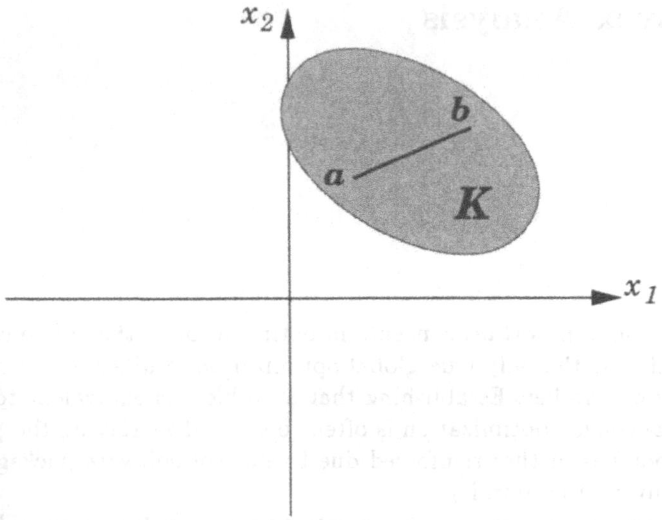

Fig. 3.1. In a convex set a chord joining any two elements of the set lies inside the set.

$$z + w = \lambda(x_1 + w) + (1 - \lambda)(x_2 + w).$$

However, $x_1 + w$ and $x_2 + w$ are elements in C. Therefore, from convexity of C, $z + w$ is in C. Thus, z is in $int(C)$. The proof that $\overline{C}$ is convex is left to the reader. □

Definition 3.1.2 (Convex combination). *A vector of the form $\sum_{k=1}^{n} \lambda_k x_k$, where $\sum_{k=1}^{n} \lambda_k = 1$ and $\lambda_k \geq 0$ for all $k = 1, \ldots, n$ is a convex combination of the vectors $x_1, \ldots, x_n$.*

Definition 3.1.3 (Cones). *A subset C of a vector space X is a cone if for every non-negative α in R and c in C, $\alpha c \in C$.*

A subset C of a vector space is a convex cone if C is convex and is also a cone.

Definition 3.1.4 (Positive cones). *A convex cone P in a vector space X is a positive convex cone if a relation '$\geq$' is defined on X based on P such that for elements x and y in X, $x \geq y$ if $x - y \in P$. We write $x > 0$ if $x \in int(P)$. Similarly $x \leq y$ if $x - y \in -P := N$ and $x < 0$ if $x \in int(N)$. Given a vector space X with positive cone P the positive cone in X^*, $P^{\oplus}$ is defined as*

$$P^{\oplus} := \{x^* \in X^* : < x, x^* > \geq 0 \text{ for all } x \in P\}.$$

Example 3.1.1. Consider the real number system R. The set

$$P := \{x : x \text{ is nonnegative}\},$$

defines a cone in R. It also induces a relation $\geq$ on R where for any two elements x and y in R, $x \geq y$ if and only if $x - y \in P$. The convex cone P with the relation $\geq$ defines a positive cone on R.

Definition 3.1.5 (Convex maps). *Let X be a vector space and Z be a vector space with positive cone P. A mapping, $G : X \to Z$ is convex if $G(tx + (1-t)y) \leq tG(x) + (1-t)G(y)$ for all x, y in X and t with $0 \leq t \leq 1$ and is strictly convex if $G(tx + (1-t)y) < tG(x) + (1-t)G(y)$ for all $x \neq y$ in X and t with $0 < t < 1$.*

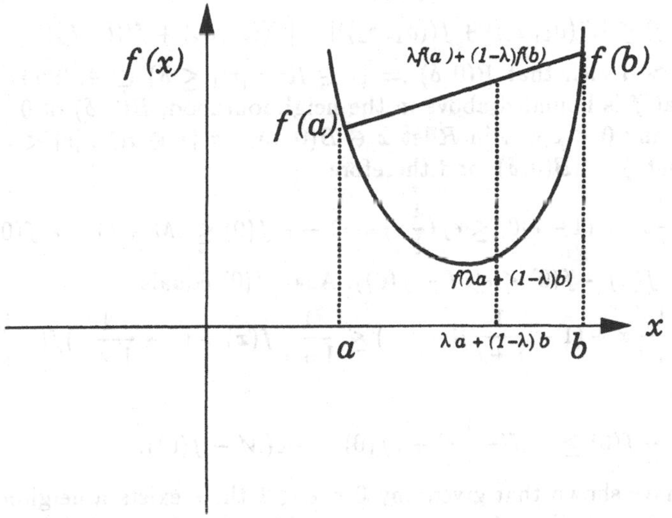

Fig. 3.2. A convex function.

Lemma 3.1.2. *Let $(X, \|\cdot\|_X)$ be a finite dimensional normed vector space and let C be a convex subset of X. Let $f : C \to R$ be a convex functional on C. Then f is continuous on $int(C)$.*

Proof. From Lemma 1.5.3, we know that it suffices to prove the lemma for $(R^n, |\cdot|_1)$. We will prove the lemma for R^2 which can be easily generalized to R^n. We will also assume that $0 \in int(C)$ (there is no loss of generality in doing so). We will show that there exists an $\epsilon > 0$ and $M \in R$ such that if $(x, y) \in R^2$, and $|(x, y)|_1 \leq \epsilon$ then $f((x, y)) \leq M$ (i.e. f is bounded above in some neighbourhood of 0.)

As $0 \in int(C)$ we know that there exists a "rectangle" A such that

$$A := \{(x, y) : a_1 \leq x \leq b_1, \text{ and } a_2 \leq y \leq b_2\},$$

where a_1, a_2, b_1 and b_2 are elements in R, with $A \subset C$. Let the sets seg_1 and seg_2 be defined as

$$\{x : x = \lambda(a_1, a_2) + (1-\lambda)(a_1, b_2) \text{ such that } \lambda \in R \text{ and } 0 \le \lambda \le 1\}$$

and

$$\{x : x = \lambda(b_1, a_2) + (1-\lambda)(b_1, b_2) \text{ such that } \lambda \in R \text{ and } 0 \le \lambda \le 1\},$$

respectively. For any element (x, y) in seg_1 we have

$$f((x,y)) \le \lambda f((a_1,a_2)) + (1-\lambda)f((a_1,b_2)) \le |f((a_1,a_2)) + f((a_1,b_2))|.$$

Similarly, for any element $(x, y) \in seg_2$, $f((x, y)) \le |f((b_1, a_2) + f((b_1, b_2))|$. It is clear that any element (x, y) of A is a convex combination of elements in seg_1 and seg_2. Thus, for all $(x, y) \in A$,

$$(x,y)) \le |f((a_1,a_2)) + f((a_1,b_2))| + |f((b_1,a_2) + f((b_1,b_2))| =: M.$$

Choose $\delta > 0$ such that $B(0, \delta) := \{r \in R^2 : |r|_1 \le \delta\} \subset A$. Thus we have shown that f is bounded above in the neighbourhood, $B(0, \delta)$ of 0 by M.

Given any $0 < \epsilon < 1$ in R let $x \in B(0, \epsilon\delta) := \{r \in R^2 : |r| < \epsilon\delta\}$. This implies that $\frac{1}{\epsilon}x \in B(0, \delta)$, and therefore

$$f(\epsilon(\frac{1}{\epsilon}x) + (1-\epsilon)0) \le \epsilon f(\frac{1}{\epsilon}x) + (1-\epsilon)f(0) \le \epsilon M + (1-\epsilon)f(0).$$

Therefore, $f(x) - f(0) \le \epsilon(M - f(0))$. Also, $f(0)$ equals

$$f(\frac{1}{1+\epsilon}x + (1 - \frac{1}{1+\epsilon})(-\frac{1}{\epsilon}x)) \le \frac{1}{1+\epsilon}f(x) + (1 - \frac{1}{1+\epsilon})f(-\frac{1}{\epsilon}x).$$

Therefore,

$$f(x) - f(0) \ge -\epsilon f(-\frac{1}{\epsilon}x) + \epsilon f(0) \ge -\epsilon(M - f(0)).$$

Thus we have shown that given any $0 < \epsilon < 1$ there exists a neighbourhood $B(0, \epsilon\delta)$ of 0 such that if x is in this neighbourhood then $|f(x) - f(0)| \le \epsilon(M - f(0))$. Therefore, f is continuous at 0.

□

3.2 Separation of Disjoint Convex Sets

Consider the vector space R^2. The equation of a line (see Figure 3.3) in R^2 is given by

$$m_1x_1 + m_2x_2 = c,$$

where m_1, m_2 and c are constants. The graph of a line is given by the set

$$L = \{(x_1, x_2) | m_1x_1 + m_2x_2 = c\},$$

which can be written as

$$L = \{x \in R^2 | < x, x^* >= c\}, \qquad (3.1)$$

where $x^* = (m_1, m_2)$. Note that if $m_2 = 0$ then we have a vertical line. We now generalize the concept of a line in R^2 to normed vector spaces.

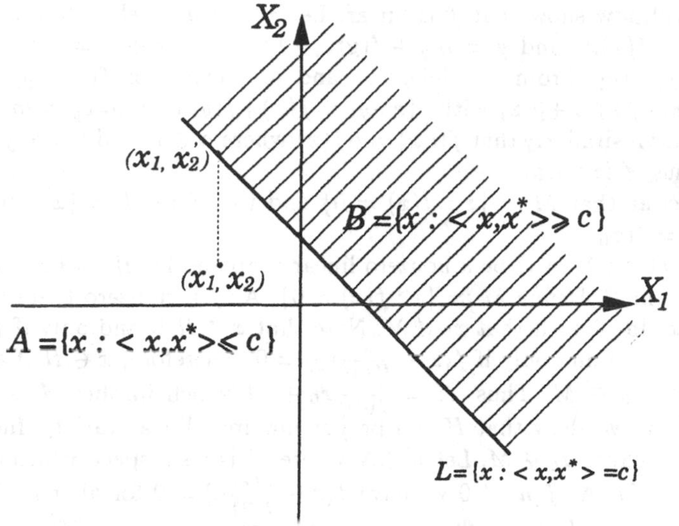

Fig. 3.3. Separation of R^2 into half spaces by a line L.

Definition 3.2.1 (Linear variety, Hyperplanes). *A subset V of a vector space X is a linear variety if there exists an element x_v in X such that*

$$V = x_v + M := \{x : x = x_v + m \text{ for some } m \in M\},$$

where M is a subspace of X.

A subset H of X is called a hyperplane if it is a linear variety which is proper (i.e. there exists an element x_0 in X which is not in H) and maximal (i.e. if H_1 is another linear variety which contains H then $H_1 = X$).

The line L defined earlier is a hyperplane in R^2.

Theorem 3.2.1. *H is a hyperplane in X if and only if there exists a nonzero linear function $f : X \to R$ such that*

$$H := \{x : f(x) = c\},$$

where $c \in R$.

Proof. ($\Rightarrow$) Let H be a hyperplane in X. Then there exists x_0 in X such that $H = x_0 + M := \{x_0 + m : m \in M\}$ where M is a proper subspace of X which is maximal. Let x_1 in X be such that $x_1 \notin M$. It is clear that the set

$$M \oplus Rx_1 := \{m + y : m \in M \text{ and } y = \alpha x_1 \text{ for some } \alpha \in R\}$$

is equal to X (because it is a subspace which contains M and M is maximal). It is also clear that for any x in X there exists unique elements $f(x) \in R$ and $m_x \in M$ such that $x = m_x + f(x)x_1$.

We will now show that f is linear. Let x and y be elements in X, then $x = m_x + f(x)x_1$ and $y = m_y + f(y)x_1$. Therefore $x + y = (m_x + m_y) + (f(x) + f(y))x_1$. From the definition and uniqueness of $f(x + y)$ we have $x+y = m_{x+y}+f(x+y)x_1$ with $f(x+y) = f(x)+f(y)$ and $m_{x+y} = m_x+m_y$. It can be shown similarly that $f(\alpha x) = \alpha f(x)$ where $\alpha \in R$ and x is any element in X. Thus, f is linear.

It is clear that $M = \{x : f(x) = 0\}$ and therefore $H = \{x : f(x) = c\}$ where $c := f(x_0)$.

($\Leftarrow$) Let $f : X \to R$ be a nonzero linear map and let $H := \{x : f(x) = c\}$ for some c in R. Define $M := \{x : f(x) = 0\}$. As f is nonzero there exists an element x_0 in X such that $x_0 \notin M$. Note that $x \in H$ if and only if $f(x) = c$ which is true if and only if $f(x - \frac{c}{f(x_0)}x_0) = 0$. Therefore, $x \in H$ if and only if $x - \frac{c}{f(x_0)}x_0 \in M$. Thus, $H = \frac{c}{f(x_0)}x_0 + M$ which implies H is a linear variety. Now we show that H is a proper maximal linear variety. Indeed, H is proper because $x_0 \notin M$. Let $n \in N$ where N is a subspace which contains M and $n \notin M$. As $f(n) \neq 0$ we have $f(x - \frac{f(x)}{f(n)}n) = 0$ for all x in X which implies that $x - \frac{f(x)}{f(n)}n \in M$ for all x in X. Thus $x = m + \frac{f(x)}{f(n)}n$ for some $m \in M$. As N is a subspace which contains M and $n \in N$ it follows that $x \in N$. As $x \in X$ was chosen arbitrarily it follows that $N = X$. Thus we have established that M is a proper maximal subspace. This proves that H is a hyperplane. □

With this theorem we have recovered the familiar description as given in equation (3.1) for hyperplanes.

Corollary 3.2.1. *Let H be a hyperplane in a vector space X such that 0 is not in H. Then there exists a unique nonzero linear map $f : X \to R$ such that $H = \{x : f(x) = 1\}$.*

Proof. From Theorem 3.2.1 it is clear that there exists a linear map $f_1 : X \to R$ and $c \in R$ such that $H = \{x : f_1(x) = c\}$. As $0 \notin H$ it is not possible that $c = 0$. Let $f : X \to R$ be defined by $f(x) = \frac{1}{c}f_1(x)$ for any x in X. Then it follows that $H = \{x : f(x) = 1\}$.

We will show that f is unique. Let $g : X \to R$ be another nonzero linear function such that $H = \{x : g(x) = 1\}$. Let h be an arbitrary element in H. Then it is clear that for any x in X, $f(x+(1-f(x))h) = 1$ which implies that $x + (1 - f(x))h \in H$. Therefore, $g(x + (1 - f(x))h) = 1$ from which it follows that $g(x) = f(x)$ for any x in X (because $g(h) = 1$). Thus, f is unique. □

For the purposes of the discussion below we will assume that c, m_1 and m_2 which describe the line L in Figure 3.1 are all nonnegative. The results for other cases will be similar. Consider the region A in Figure 3.1 which is the region "below" the line L. As illustrated earlier, $L = \{x :< x, x^* >= c\}$ where $x^* = (m_1, m_2)$. Consider any point $x = (x_1, x_2)$ in region A. Such a point lies "below" the line L. Thus if $x' = (x_1, x_2')$ denotes the point on the line L which has the same first coordinate as that of x then $x_2' \geq x_2$. As x' is on the line L it follows that $< x', x^* >= m_1x_1 + m_2x_2' = c$. As $m_2 \geq 0$ it follows that $< x, x^* >= m_1 x_1 + m_2 x_2 \leq m_1 x_1 + m_2 x_2' = c$. Thus we have shown that for every point x in the region A, $< x, x^* > \leq c$. In a similar manner it can be established that if $< x, x^* > \leq c$ then x lies "below" the line L, that is $x \in A$. Thus the region A is given by the set $\{x :< x, x^* > \leq c\}$ which is termed the negative half space of L. In an analogous manner it can be shown that the region B (which is the region "above" the line L) is described by $\{x :< x, x^* > \geq c\}$. This set is termed the positive half space of L. Thus the line L separates R^2 into two halves; a positive and a negative half. We generalize the concept of half spaces for an arbitrary normed vector space.

Definition 3.2.2 (Half spaces). *Let $(X, \|\cdot\|_X)$ be a normed linear space and let $x^* : X \to R$ be a bounded linear functon on X. Let*

$$\begin{aligned} S_1 &:= \{x \in X :< x, x^* > < c\}, \\ S_2 &:= \{x \in X :< x, x^* > \leq c\}, \\ S_3 &= \{x \in X :< x, x^* > > c\}, \\ S_4 &:= \{x \in X :< x, x^* > \geq c\}. \end{aligned}$$

Then S_1 is an open negative half space, S_2 is a closed negative half space, S_3 is an open positive half space and S_4 is a closed positive half space.

Lemma 3.2.1. *Let $(X, \|\cdot\|_X)$ be a normed linear space and let $f : X \to R$ be a bounded linear functon on X. Then the sets $\{x \in X : f(x) < c\}$ and $\{x \in X : f(x) > c\}$ are open and the sets $\{x \in X : f(x) \leq c\}$ and $\{x \in X : f(x) \geq c\}$ are closed in the norm topology for any c in R.*

Proof. The proof is left to the reader. □

It is intuitively clear that in R^2 if two convex sets C_1 and C_2 do not intersect then there exists a line in R^2 which separates the two sets (see Figure 3.4). In other words there exists x^* in $(R^2)^*$ and a constant c in R such that C_1 lies on the positive half space of the line $L = \{x | < x, x^* >= c\}$ and C_2 lies in the negative half space of L. That is

$$C_1 \subset \{x :< x, x^* > \geq c\},$$

and

$$C_2 \subset \{x :< x, x^* > \leq c\}.$$

The main focus of this section is to generalize this result to disjoint convex sets in a general normed vector space. In this regard we can immediately establish the following result. Suppose $(X, \|.\|_X)$ is a normed vector space and let $B := \{x | \|x\| \leq 1\}$ be the unit norm ball such that $x_0 \notin int(B)$. Then it is possible to separate B and x_0 by a hyperplane. Indeed from Theorem 2.2.4, it follows that there exists x^* in X^* such that $\|x^*\| \leq 1$ and $< x_0, x^* >= \|x_0\|_X \geq 1$. As $\|x^*\| \leq 1$ and $\|x\|_X < 1$ it follows that

$$< x, x^* > \leq \|x^*\| \|x\|_X \leq \|x\|_X < 1 \leq \|x_0\|_X \text{ for all } x \in int(B).$$

Thus

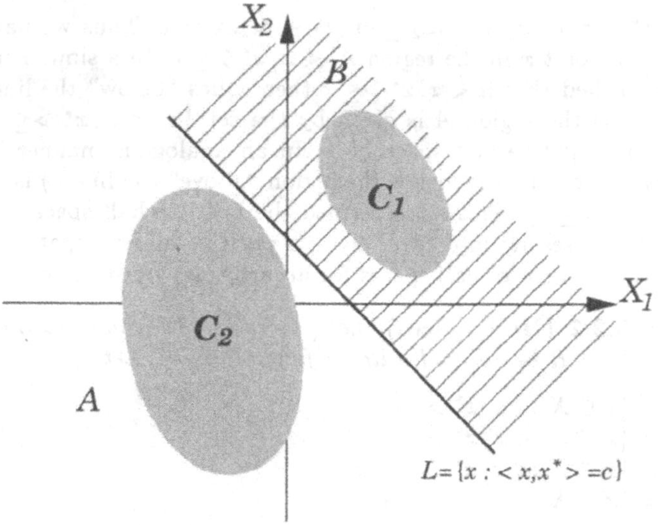

Fig. 3.4. Separation of of convex sets in R^2.

$$int(B) \subset \{x :< x, x^* >< \|x_0\|_X\},$$

whereas

$$x_0 \in L := \{x :< x, x^* >= \|x_0\|_X\}.$$

Thus we have shown that it is possible to separate the interior of a unit norm ball and any element which does not belong to the interior of the unit norm ball. Minkowski's function is a norm like functional associated with a convex set which allows us to use a similar argument as developed above to separate disjoint convex sets.

Lemma 3.2.2 (Minkowski's function). *Let K be a convex subset of a normed linear space $(X, \|\cdot\|_X)$ such that $0 \in int(K)$. For any $x \in X$ let*

$$p(x) := \inf\{\lambda \in R : \lambda > 0, \text{ such that } x \in \lambda K\},$$

where

$$\lambda K := \{x : X = \lambda k \text{ for some } k \in K\}.$$

Then p is a real valued continuous sublinear function which is non-negative.

Proof. It is clear that p is non-negative. As $0 \in int(K)$, there exists an $a > 0$ in R such that if for any $k \in X$, $\|k\|_X < a$ implies that $k \in K$. Let x be any element in X such that $\|x\|_X \neq 0$. Then $\|\frac{ax}{2\|x\|_X}\|_X < a$ and therefore $x \in \frac{2\|x\|_X}{a}K$. Thus, for any $x \in X$, $p(x) < \frac{2\|x\|_X}{a} < \infty$. Thus we have shown that p is real valued and non-negative.

Let $\alpha \in R$ be such that $\alpha > 0$. Then

$$\begin{aligned} p(\alpha x) &= \inf\{\lambda \in R : \lambda > 0 \text{ and } \alpha x \in \lambda K\} \\ &= \inf\{\alpha \tfrac{\lambda}{\alpha} \in R : \tfrac{\lambda}{\alpha} > 0 \text{ and } x \in \tfrac{\lambda}{\alpha} K\} \\ &= \alpha p(x). \end{aligned}$$

As $0 \in int(K)$ it follows that $p(0) = 0$. Let x and y be elements in X. Note that if $p(z) < 1$ then $z \in K$. Indeed, if $p(z) < 1$ then there exists $\lambda \in R$ and $k \in K$ such that $0 < \lambda < 1$ with $z = \lambda k = \lambda k + (1-\lambda)0$. As k and 0 both are in K and K is convex it follows that z is in K. Given any $\epsilon > 0$ let $r_x \in R$ and $r_y \in R$ be such that $p(x) < r_x < p(x) + \frac{\epsilon}{2}$ and $p(y) < r_y < p(y) + \frac{\epsilon}{2}$. As, $1 > \frac{1}{r_x} p(x) = p(\frac{x}{r_x})$ we know that $\frac{x}{r_x} \in K$. Similarly, $\frac{y}{r_y} \in K$. Let $r := r_x + r_y$. From the convexity of K, $\frac{r_x}{r}\frac{x}{r_x} + \frac{r_y}{r}\frac{y}{r_y} \in K$. This implies that $\frac{1}{r}(x+y) \in K$ and therefore $x + y \in rK$. Thus, $p(x+y) \leq r \leq p(x) + p(y) + \epsilon$. As $\epsilon > 0$ is arbitrary it follows that $p(x+y) \leq p(x) + p(y)$. This proves that p is sublinear.

Note that for elements x and y in X, $p(x) = p(x-y+y) \leq p(x-y) + p(y)$. This implies that $p(x) - p(y) \leq p(x-y) \leq \frac{2\|x-y\|_X}{a}$. Similarly it follows that $p(y) - p(x) \leq \frac{2\|y-x\|_X}{a}$ and therefore $|p(x) - p(y)| \leq \frac{2\|x-y\|_X}{a}$. Thus p is a continuous function. □

Theorem 3.2.2. *Let $(X, \|\cdot\|_X)$ be a normed linear space and let X^* denote its dual. Let K be a convex subset of X such that $int(K) \neq \{\}$ and let V be a linear variety such that $int(K) \cap V = \{\}$. Then there exists a nonzero x^* in X^* such that*

$$\begin{aligned} int(K) &\subset \{x \in X : \langle x, x^* \rangle < c\} \\ \overline{K} &\subset \{x \in X : \langle x, x^* \rangle \leq c\} \\ V &\subset \{x \in X : \langle x, x^* \rangle = c\}, \end{aligned}$$

where $\overline{K}$ denotes the closure of the set K in the norm topology.

Proof. We will first prove the theorem when $0 \in int(K)$. Let $V = x_0 + N$ where x_0 is an element in X and N is a subspace of X. Let

$$M = N \oplus Rx_0 := \{n + y : n \in N \text{ and } y = \alpha x_0 \text{ for some } \alpha \in R\}.$$

For any element m in M let $m = n_m + f(m)x_0$. We now show that n_m and $f(x)$ are unique. Let $m = n_1 + \alpha x_0 = n_2 + \beta x_0$ where α and β in R with $\alpha \neq \beta$ and n_1 and n_2 are elements in M. Then it follows that $x_0 = \frac{n_2 - n_1}{\alpha - \beta}$ and therefore x_0 is in N. As N is a subspace we have $-x_0 \in N$. This implies that $0 \in V$ which is not true because $int(K) \cap V = \{\}$ and $0 \in int(K)$. Thus n_m and $f(m)$ are unique for every m in M. Thus f defines a function on M. It can also be shown that f is a linear function. Note that

$$V = \{m \in M : f(m) = 1\}.$$

For any x in X let

$$p(x) = \inf\{\lambda \in R : x \in \lambda K\}.$$

p is the Minkowski's function of the convex set K. As $int(K) \cap V = \{\}$ it follows that for all $v \in V$, $p(v) \geq 1$. Therefore, if v is in M and $f(v) = 1$ then $p(v) \geq 1$. For all $m \in M$, with $f(m) > 0$, $f(\frac{m}{f(m)}) = 1$ which implies that for all m in M with $f(m) > 0$, $p(\frac{m}{f(m)}) \geq 1$. From the sublinearity and the non-negativity of p (see Lemma 3.2.2) it follows that $p(m) \geq f(m)$ for all m in M.

From Theorem 2.1.2 we know that there exists a linear function $F : X \to R$ such that $F(m) = f(m)$ for all m in M and $F(x) \leq p(x)$ for all x in X.

It is clear that F is continuous because p is continuous and $F(x) \leq p(x)$ for all x in X. We rename F as x^*. Then it is clear that for every element k in $int(K)$, $p(k) < 1$ and therefore $< k, x^* > < 1$. Also, for every element k in $\overline{K}$, $< k, x^* > \leq p(k) \leq 1$. As $F(v) = f(v) = 1$ for every element v in V it follows that $< v, x^* > = 1$ on V. Also, $x^* \neq 0$ as V is not empty. Thus we have established the theorem when $0 \in int(K)$.

For the more general case if k_0 is in $int(K)$ then let

$$K' := \{k - k_0 : k \in K\}, \text{ and let } V' := \{v - k_0 : v \in V\}.$$

With these definitions we have that $0 \in int(K')$ and $V' \cap int(K') = \{\}$ (as $V \cap int(K) = \{\}$). Applying the result to K' and V' the theorem in the general case follows easily with $c :=< k_0, x^* >$. □

Corollary 3.2.2 (Separation of a point and a convex set). *Let K be a convex subset of a normed linear space $(X, \|\cdot\|_X)$ with $int(K) \neq \{\}$. Let x_0 in X be such that $x_0 \notin int(K)$. Then there exists a nonzero $x^* \in X^*$ such that*

$$< k, x^* > < < x_0, x^* > \text{ for all } k \text{ in } int(K) \text{ and}$$
$$< k, x^* > \leq < x_0, x^* > \text{ for all } k \text{ in } \overline{K}.$$

Proof. Let $V := \{x_0\}$. Then V is a linear variety such that $V \cap int(K) = \{\}$. From Theorem 3.2.2 it follows that there exists x^* in X^* and $c \in R$ such that $< x_0, x^* >= c$, $< k, x^* > < c$ for all k in $int(K)$ and $< k, x^* > \leq c$ for all k in $\overline{K}$. This proves the corollary. □

Corollary 3.2.3. *Let K be a convex subset of a normed linear space $(X, \|\cdot\|_X)$ with $int(K) \neq \{\}$. Let $x_0 \in bd(K) := \overline{K} \setminus int(K) := \{x \in \overline{K} : x \notin int(K)\}$. Then there exists a nonzero $x^* \in X^*$ such that $< x_0, x^* >= \sup\{< k, x^* >: k \in K\}$.*

Proof. From Corollary 3.2.2 we know that there exists x^* in X^* such that

$$< k, x^* > \leq < x_0, x^* > \text{ for all } k \in K. \tag{3.2}$$

As $x_0 \in \overline{K}$ we know that there exists a sequence $\{x_n\}$ in K such that $\|x_0 - x_n\|_X \to 0$. From continuity of x^* it follows that $< x_n, x^* > \to < x_0, x^* >$. From equation 3.2 we conclude that $< x_0, x^* > = \sup\{< k, x^* >: k \in K\}$. □

The following corollary is often referred to as the Eidelheit separation result.

Corollary 3.2.4 (Separation of disjoint convex sets). *Let K_1 and K_2 be convex subsets of a normed linear space $(X, \|\cdot\|_X)$. Let $int(K_1) \neq \{\}$ and suppose $int(K_1) \cap K_2 = \{\}$. Then there exists a nonzero x^* in X^* such that*

$$\sup\{< x, x^* >: x \in K_1\} \leq \inf\{< x, x^* >: x \in K_2\}.$$

Proof. Let $K := K_1 - K_2 := \{k_1 - k_2 : k_1 \in K_1 \text{ and } k_2 \in K_2\}$. As $int(K_1) \cap K_2 = \{\}$ it follows that $0 \notin int(K)$. Also, $int(K) \neq \{\}$ because $int(K_1) \cap K_2 = \{\}$ and $int(K_1) \neq \{\}$. From Corollary 3.2.2 there exists a nonzero x^* in X^* such that $< k, x^* > \leq 0$ for all k in K. This implies that for any k_1 in K_1 and for any k_2 in K_2, $< k_1, x^* > \leq < k_2, x^* >$. This proves the corollary. □

3.3 Convex Optimization

The problem that is the subject of the rest of the chapter is the following problem.

$$\mu = \inf f(x) \quad \text{subject to} \quad x \in \Omega,$$

where $f : \Omega \to R$ is a convex function on a convex subset Ω of a vector space X. Such a problem is called a convex optimization problem.

Lemma 3.3.1. *Let $f : (X, \|.\|_X) \to R$ be a convex function and let ω be a convex subset of X. If there exists a neighbourhood N in Ω of ω_0 where $\omega_0 \in \Omega$ such that for all $\omega \in N$, $f(\omega_0) \leq f(\omega)$ then $f(\omega_0) \leq f(\omega)$ for all ω in Ω (that is every local minimum is a global minimum).*

Proof. Let ω be any element of Ω. Let $0 \leq \lambda \leq 1$ be such that $x := \lambda\omega_0 + (1-\lambda)\omega$ be in N. Then $f(\omega_0) \leq f(x) \leq \lambda f(\omega_0) + (1-\lambda)f(\omega)$. This implies that $f(\omega_0) \leq f(\omega)$. As ω is an arbitrary element of Ω we have established the lemma. □

Lemma 3.3.2. *Let Ω be a convex subset of a Banach space X and $f : \Omega \to R$ be strictly convex. If there exists an $x_0 \in \Omega$ such that*

$$f(x_0) = \inf_{x \in \Omega} f(x),$$

(that is f achieves its minimum on Ω) then the minimizer is unique.

Proof. Let $m := \min_{x\in\Omega} f(x)$. Let $x_1, x_2 \in \Omega$ be such that $f(x_1) = f(x_2) = m$. Let $0 < \lambda < 1$. From convexity of Ω we have $\lambda x_1 + (1-\lambda)x_2 \in \Omega$. From strict convexity of f we have that if $x_1 \neq x_2$ then $f(\lambda x_1 + (1-\lambda)x_2) < \lambda f(x_1) + (1-\lambda)f(x_2) = m$ which is a contradiction. Therefore $x_1 = x_2$. This proves the lemma. □

Many convex optimization problems have the following structure

$$\begin{aligned} \omega(z) = \quad & \inf f(x) \\ \text{subject to} \quad & \\ & x \in \Omega \\ & g(x) \leq z, \end{aligned} \tag{3.3}$$

where $f : \Omega \to R$, $g : X \to Z$ are convex maps with Ω a convex subset of the vector space X and Z a normed vector space with a positive cone P. The condition $g(x) \leq z$ is to be interpreted with respect to the positive cone P of the vector space Z.

Lemma 3.3.3. *The function ω is convex.*

Proof. Let z_1 and z_2 be elements in Z and let $0 \leq \lambda \leq 1$ be any constant. Then

$$\begin{aligned} \omega(\lambda z_1 + (1-\lambda)z_2) &= \inf\{f(x) : x \in \Omega, g(x) \leq \lambda z_1 + (1-\lambda)z_2\} \\ &= \inf\{f(x) : x = \lambda x_1 + (1-\lambda)x_2, x_1 \in \Omega, x_2 \in \Omega, \\ &\qquad g(x) \leq \lambda z_1 + (1-\lambda)z_2\} \\ &\leq \inf\{\lambda f(x_1) + (1-\lambda)f(x_2), x_1 \in \Omega, x_2 \in \Omega, \\ &\qquad g(x) \leq \lambda z_1 + (1-\lambda)z_2\} \\ &\leq \inf\{\lambda f(x_1) + (1-\lambda)f(x_2), x_1 \in \Omega, x_2 \in \Omega, \\ &\qquad g(x_1) \leq z_1, g(x_2) \leq z_2\} \\ &= \lambda\omega(z_1) + (1-\lambda)\omega(z_2). \end{aligned}$$

The second equality is true because for any given λ with $0 \leq \lambda \leq 1$ the set $\Omega = \{x : x = \lambda x_1 + (1-\lambda)x_2, x_1 \in \Omega, x_2 \in \Omega\}$. The first inequality is true because f is a convex map. The second inequality is true because the set $\{(x_1, x_2) \in \Omega \times \Omega : g(\lambda x_1 + (1-\lambda)x_2) \leq \lambda z_1 + (1-\lambda)z_2\} \supset \{(x_1, x_2) \in \Omega \times \Omega : g(x_1) \leq z_1, g(x_2) \leq z_2\}$, which follows from the convexity of g. This proves the lemma. □

Lemma 3.3.4. *Let z_1 and z_2 be elements in Z such that $z_1 \leq z_2$ with respect to the convex cone P. Then $\omega(z_2) \leq \omega(z_1)$.*

Proof. Follows immediately from the relation $\{x \in \Omega : g(x) \leq z_2\} \supset \{x \in \Omega : g(x) \leq z_1\}$, if $z_1 \leq z_2$. □

Definition 3.3.1 (Epigraph). *Let $f : \Omega \to R$ be a real valued function where Ω is a subset of a vector space X. The epigraph of f over Ω is a subset $[f, \Omega]$ of $R \times X$ defined by*

$$[f, \Omega] := \{(r, \omega) \in R \times X : x \in \Omega, f(x) \leq r\}.$$

Lemma 3.3.5. *Let $f : \Omega \to R$ be a real valued function where Ω is a convex subset of a vector space X. Then f is convex if and only if $[f, \Omega]$ is convex.*

Proof. Left to the reader. □

3.3.1 Minimum Distance to a Convex Set

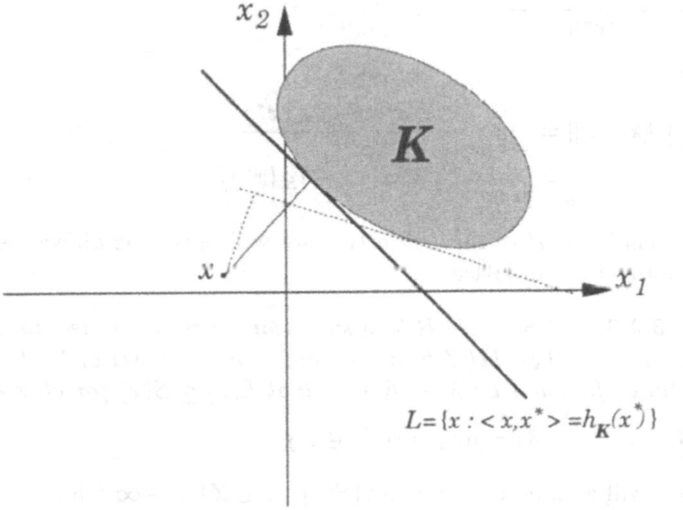

Fig. 3.5. Support hyperplane to a convex set K. The figure also illustrates the fact that the minimum distance from a point x to a convex set K is the maximum of the distances of the point from the supporting hyperplanes of the convex set.

Note that the minimum distance of a point $x = (x_1, x_2)$ from a line (see Figure 3.3) in R^2 given by

$$\inf_{y \in L} ||y - x||,$$

is equal to

$$\frac{m_1 x_1 + m_2 x_2 - c}{\sqrt{m_1^2 + m_2^2}} = \frac{< x, x^* > -c}{||x^*||},$$

where the equation of the line is given by $m_1 x_1 + m_2 x_2 = c$ and $x^* = (m_1, m_2)$.

Definition 3.3.2 (Support-functional). *Let K be a nonempty convex subset of a normed linear space $(X, || \cdot ||_X)$. The support functional $h : X^* \to R \cup \{\infty\}$ is defined by*

$$h_K(x^*) := \sup\{< k, x^* >: k \in K\},$$

for any x^* in X^*.

Note that for all $k \in K$, $< k, x^* > \leq h_K(x^*)$. Thus it is clear that the hyperplane $L = \{< x, x^* >= h_k(x^*)\}$ divides the vector space into two halves such that the convex set K lies entirely in one half of the space (see Figure 3.5). By inspection it can be seen that in R^2, the minimum distance of a point x from a convex set K is equal to the maximum of the distances of the point x from the supporting hyperplanes. As the distance of the point x from the supporting hyperplane associated with $h_K(x^*)$ is given by

$$\frac{< x, x^* > -h_K(x^*)}{||x^*||},$$

we have

$$\begin{aligned}\inf_{k \in K} ||k - x|| &= \max_{x^* \in R^2} \frac{< x, x^* > -h_K(x^*)}{||x^*||} \\ &= \max_{||x^*|| \leq 1} \{< x, x^* > -h_K(x^*)\}.\end{aligned}$$

In the remaining part of this subsection we will prove the above result for a general normed vector space.

Lemma 3.3.6. *Let $S : X \to R$ be a sublinear function defined on a normed linear space $(X, ||\cdot||_X)$. Let Z be a nonempty convex subset of X. Then, there exists a linear function $L : X \to R$ such that $L(x) \leq S(x)$ for all $x \in X$ and*

$$\inf\{L(z) : z \in Z\} = \inf\{S(z) : z \in Z\}.$$

Proof. We will assume that $a := \inf\{S(z) : z \in Z\} > -\infty$ (the proof for the case when $a = -\infty$ follows easily from Theorem 2.1.1). For any x in X let

$$U(x) := \inf\{S(x + \lambda z) - \lambda a : z \in Z, \lambda \geq 0\}.$$

We will show that U is a real valued sublinear function such that $U(x) \leq S(x)$ for all x in X and for all $z \in Z$, $U(-z) \leq -a$. Let x and y be elements in X, then

$$\begin{aligned}U(x) + U(y) &= \inf\{S(x + \lambda_1 z_1) + S(y + \lambda_2 z_2) \\ &\qquad -(\lambda_1 + \lambda_2)a : z_1, z_2 \in Z, \lambda_1, \lambda_2 \geq 0\} \\ &\geq \inf\{S(x + y + \lambda_1 z_1 + \lambda_2 z_2) - (\lambda_1 + \lambda_2)a \\ &\qquad : z_1, z_2 \in Z, \lambda_1, \lambda_2 \geq 0\} \\ &= \inf\{S(x + y + (\lambda_1 + \lambda_2)(\tfrac{\lambda_1}{\lambda_1+\lambda_2} z_1 + \tfrac{\lambda_2}{\lambda_1+\lambda_2} z_2)) \\ &\qquad -(\lambda_1 + \lambda_2)a : z_1, z_2 \in Z, \lambda_1, \lambda_2 \geq 0\} \\ &= \inf\{S(x + y + \lambda z) - \lambda a : z \in Z, \lambda \geq 0\} = U(x + y).\end{aligned}$$

The third equality is true because for $\lambda_1, \lambda_2 \geq 0$, $\{\frac{\lambda_1}{\lambda_1+\lambda_2} z_1 + \frac{\lambda_2}{\lambda_1+\lambda_2} z_2 : z_1, z_2 \text{ in } Z\} = Z$ and the set $\{\lambda_1 + \lambda_2 : \lambda_1, \lambda_2 \geq 0\} = \{\lambda : \lambda \geq 0\}$.

Let $\alpha > 0$, then

$$\begin{aligned}U(\alpha x) &= \inf\{S(\alpha x) - \lambda a : z \in Z, \lambda \geq 0\} \\ &= \alpha \inf\{S(x) - \tfrac{\lambda}{\alpha} a : z \in Z, \lambda \geq 0\} \\ &= \alpha \inf\{S(x) - \hat{\lambda} a : z \in Z, \lambda \geq 0\}.\end{aligned}$$

The second equality is true because $S(\alpha x) = \alpha S(x)$ and the last equality is true because $\{\frac{\lambda}{\alpha} : \lambda \geq 0\} = \{\lambda : \lambda \geq 0\}$. It is clear that $U(0) = 0$. Thus, we have established that $U(\alpha x) = \alpha U(x)$ for all $x \in X$ and for all $\alpha \geq 0$.
Note that for any x in X, $0 = U(0) = U(x - x) \leq U(x) + U(-x) \leq U(x) + S(-x)$ which implies that $U(x) \geq -S(-x) > -\infty$. Also it follows easily from the definition of U that $U(x) \leq S(x) < \infty$. Therefore, we have shown that U is a real valued sublinear function such that for all x in X, $U(x) \leq S(x)$.

From Theorem 2.1.1 we know that there exists a linear function $L : X \to R$ such that for all x in X, $L(x) \leq U(x) \leq S(x)$. This implies that $\inf\{L(z) : z \in Z\} \leq \inf\{S(z) : z \in Z\} = a$.

Note that for any $w \in Z$

$$U(-w) = \inf\{S(-w + \lambda z) - \lambda a : \lambda \geq 0, z \in Z\} \leq S(w - w) - a = -a.$$

Therefore, for any $z \in Z$, $L(-z) \leq U(-z) \leq -a$ which implies that for all z in Z $L(z) \geq a$ (because L is linear). Thus, we have $\inf\{L(z) : z \in Z\} \leq \inf\{S(z) : z \in Z\} = a$ and therefore $\inf\{L(z) : z \in Z\} = a$. This proves the lemma. ∎

Theorem 3.3.1 (Minimum distance from a convex set). *Let K be a nonempty convex subset of a normed space $(X, ||\cdot||_X)$ and let x_0 be an element in X. Let the dual space of $(X, ||\cdot||_X)$ be denoted by X^*. Define*

$$\mu := \inf\{||x_0 - k||_X : k \in K\}.$$

Then,

$$\mu = \max\{< x_0, x^* > -h_K(x^*) : ||x^*|| \leq 1\}.$$

Also, if k_0 in K and $x_0^* \in X^*$ are such that $||x_0 - k_0||_X =< x_0, x_0^* > -h_K(x_0^*) = \mu$ then

$$< x_0 - k_0, x_0^* >= ||x_0 - k_0||_X \, ||x_0^*||.$$

Proof. We will first show that $\sup\{< x_0, x^* > -h_K(x^*) : ||x^*|| \leq 1\} \geq \mu$. Let $S : X \to R$ be a real valued sublinear function defined by $S(x) = ||x||_X$ for any x in X. Let

$$Z := x_0 - K := \{y : \text{ there exists } k \in K \text{ with } y = x_0 - k\}.$$

From Lemma 3.3.6 we know that there exists x_0^* in X^* such that $||x_0^*|| \leq 1$ and

$$\begin{aligned} \inf\{< z, x_0^* >: z \in Z\} &= \inf\{||z||_X : z \in Z\} \\ &= \inf\{||x_0 - k||_X : k \in K\} = \mu. \end{aligned}$$

However,

$$\begin{aligned} \inf\{< z, x_0^* >: z \in Z\} &= \inf\{< x_0 - k, x_0^* >: k \in K\} \\ &= < x_0, x_0^* > + \inf\{- < k, x_0^* >: k \in K\} \\ &= < x_0, x_0^* > -h_K(x_0^*). \end{aligned}$$

Thus we have established that there exists x_0^* in X^* such that $||x_0^*|| \leq 1$ and $\mu = < x_0, x_0^* > -h_K(x_0^*)$. This implies that

$$\sup\{< x_0, x^* > -h_K(x^*) : ||x^*|| \leq 1\} \geq \mu.$$

Now we will show that $\mu \geq \sup\{< x_0, x^* > -h_K(x^*) : ||x^*|| \leq 1\}$. Let x^* in X^* be such that $||x^*|| \leq 1$. Then for any $k \in K$,

$$||x_0 - k||_X \geq | < x_0 - k, x^* > | \geq < x_0, x^* > - < k, x^* > .$$

Therefore,

$$\inf\{||x_0 - k||_X : k \in K\} \geq < x_0, x^* > -h_K(x^*).$$

This holds for any x^* in X^* which satisfies $||x^*|| \leq 1$. Thus

$$\mu \geq \sup\{< x_0, x^* > -h_K(x^*) : ||x^*|| \leq 1\}.$$

However, we have established earlier that there exists $x_0^* \in X^*$ such that $||x_0^*|| \leq 1$, and $\mu = < x_0, x_0^* > -h_K(x_0^*)$. Therefore,

$$\mu = \max\{< x_0, x^* > -h_K(x^*) : ||x^*|| \leq 1\},$$

where we have replaced the term sup by max in the right hand side of the equation.

Let k_0 in K and x_0^* in X^* be such that $||x_0^*|| \leq 1$ and

$$\mu = ||x_0 - k||_X = < x_0, x_0^* > -h_K(x_0^*).$$

From the definition of h_K we have that $< k_0, x_0^* > \leq h_K(x_0^*)$ which implies that

$$< x_0 - k_0, x_0^* > \geq < x_0, x_0^* > -h_K(x_0^*) = \mu = ||x_0 - k_0||_X.$$

As $||x_0^*|| \leq 1$ it follows that $< x_0 - k_0, x_0^* > \geq ||x_0 - k_0||_X \; ||x^*||$. Thus $< x_0 - k_0, x_0^* > = ||x_0 - k_0||_X \; ||x^*||$. This proves the theorem. □

3.3.2 Kuhn-Tucker Theorem

Consider the convex optimization problem

$$\begin{aligned} \omega(z) = \quad & \inf f(x) \\ \text{subject to} \quad & \\ & x \in \Omega \\ & g(x) \leq z. \end{aligned}$$

We will obtain information about $\omega(0)$ by analyzing $\omega(z)$. We have shown that $\omega(z)$ is a decreasing function of z (see Lemma 3.3.4) and that it is a convex function (see Lemma 3.3.3). It can be visualized as illustrated in Figure 3.6. As $\omega(z)$ is a decreasing function it is evident that the tangent to the curve at $(0, \omega(0))$ has a negative slope (see Figure 3.6). Thus the tangent can be characterized by a line L with the equation:

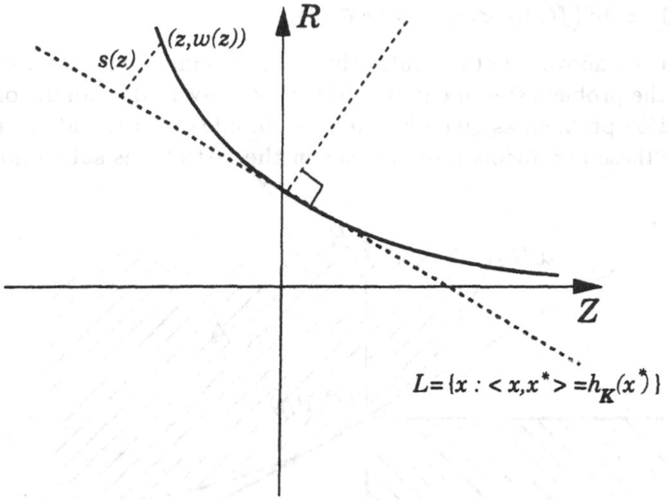

Fig. 3.6. Illustration of $\omega(z)$

$$\omega(z) + < z, z^* > = c,$$

where $z^* \geq 0$. Also, note that if we change the coordinates such that L becomes the horizontal axis and its perpendicular the vertical axis with the origin at $(0, \omega(0))$ (see Figure 3.6) then the function $\omega(z)$ achieves its minimum at the new origin. In the new cordinate system the vertical cordinate of the curve $\omega(z)$ is given by the distance of $(z, \omega(z))$ from the line L. This distance is given by

$$s(z) = \frac{\omega(z) + < z, z^* > - c}{\|(1, z^*)\|}.$$

Thus $s(z)$ achieves its minimum at $z = 0$. This implies that

$$\begin{aligned}
\omega(0) &= \min_{z \in Z} \{\omega(z) + < z, z^* >\} \\
&= \min_{z \in Z} \{\inf\{f(x) : x \in \Omega, g(x) \leq z\} + < z, z^* >\} \\
&= \inf\{f(x) + < z, z^* >: x \in \Omega, z \in Z, g(x) \leq z\} \\
&\geq \inf\{f(x) + < g(x), z^* >: x \in \Omega, z \in Z, g(x) \leq z\} \\
&\geq \inf\{f(x) + < g(x), z^* >: x \in \Omega\}.
\end{aligned}$$

The first inequality is true because $z^* \geq 0$ and $g(x) \leq z$. The second inequality is true because the $\{x \in \Omega : z \in Z, g(x) \leq z\} \subset \{x \in \Omega\}$. It is also true that

$$\begin{aligned}
\omega(0) &= \inf\{f(x) + < z, z^* >: x \in \Omega, z \in Z, g(x) \leq z\} \\
&\leq \inf\{f(x) + < z, z^* >: x \in \Omega\},
\end{aligned}$$

because $g(x) \leq g(x)$ is true for every $x \in \Omega$. Thus we have

$$\omega(0) = \inf\{f(x) + < z, z^* >: x \in \Omega\}.$$

Note that the above equation states that a constrained optimization problem given by the problem statement of $\omega(0)$ can be converted to an unconstrained optimization problem as given by the right hand side of the above equation. We make these arguments more precise in the rest of this subsection.

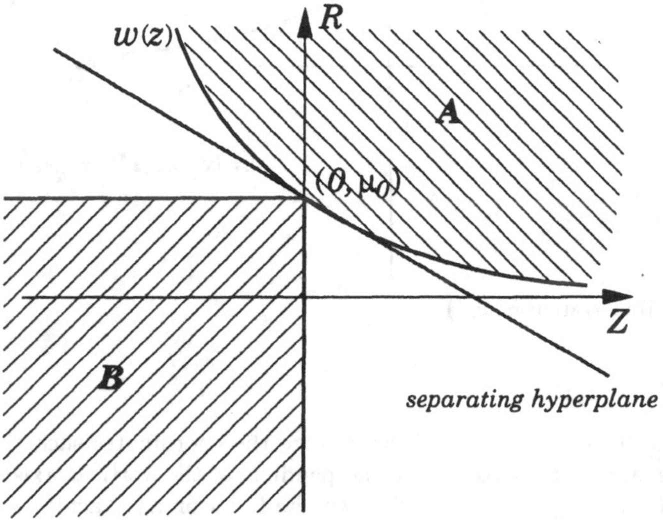

Fig. 3.7. Figure for Lemma 3.3.7.

Lemma 3.3.7. *Let $(X, ||\cdot||_X)$, and $(Z, ||\cdot||_Z)$, be normed vector spaces with Ω a convex subset of X. Let P be a positive convex cone defined in Z. Let Z^* denote the dual space of Z with the postive cone $P^\oplus$ associated with P. Let $f : \Omega \to R$ be a real valued convex functional and $g : X \to Z$ be a convex mapping. Define*

$$\mu_0 := \inf\{f(x) : \ g(x) \leq 0, \ x \in \Omega\}. \tag{3.4}$$

Suppose there exists $x_1 \in \Omega$ such that $g(x_1) < 0$ and and suppose μ_0 is finite. Then, there exist $z_0^* \geq 0$ such that

$$\mu_0 = \inf\{f(x) + < g(x), z_0^* >: x \in \Omega\}. \tag{3.5}$$

Furthermore, if there exists x_0 such that $g(x_0) \leq 0$ and $\mu_0 = f(x_0)$ then

$$< g(x_0), z_0^* >= 0 \tag{3.6}$$

Proof. We wiil say that an element x in Ω is feasible if $g(x) \leq 0$. Define A, (see Figure 3.7) a subset of $Z \times R$ by

$$A := \{(z, r) : \text{ there exists } x \in \Omega \text{ such that } g(x) \leq z \text{ and } f(x) \leq r\},$$

and B (see Figure 3.7) another subset of $Z \times R$ by

$$B := -P \times (-\infty, \mu_0] := \{(z,r) : -z \in P \text{ and } r \leq \mu_0\}.$$

We will assume that the norm on $Z \times R$ is the product norm induced by the norms on Z and R. Note that in this norm $int(B) \neq \{\}$ (let $p_0 \in int(-P)$; then $(p_0, \mu_0 - 1) \in int(B)$). We will show that $int(B) \cap A = \{\}$.

Suppose $(z, r) \in int(B) \cap A$. Then there exists x in Ω such that $f(x) \leq r$ and $g(x) \leq z$. Also $z \in -P$ and $r < \mu_0$. Therefore, $f(x) \leq r < \mu_0$ and $g(x) \leq z \leq 0$. This implies that x is feasible and $f(x)$ is strictly less than μ_0 which contradicts the definition of μ_0. Therefore, $int(B) \cap A = \{\}$.

Applying Eidelheit's separation result (see Corollary 3.2.4) to A and B (note that A and B are convex) we know that there exists a nonzero element $(z^*, s) \in (Z \times R)^* = Z^* \times R$ (see Theorem 2.2.3) and $k \in R$ such that

$$< z, z^* > +sr \geq k \text{ for all } (z,r) \in A \text{ and} \tag{3.7}$$

$$< z, z^* > +sr \leq k \text{ for all } (z,r) \in B. \tag{3.8}$$

We will now show that $s \geq 0$. As $(0, r)$ for $r \leq \mu_0$ is in B it follows from inequality (3.8) that $sr \leq k$ for all $r \leq \mu_0$. This implies that $s \geq 0$ (otherwise by letting $r \to -\infty$ we see that $k = \infty$ which is not possible because inequality (3.7) holds).

We will now show that $s > 0$. Suppose that $s = 0$. Then from inequality (3.7) we have

$$< g(x_1), z^* > \geq k, \tag{3.9}$$

because $(g(x_1), f(x_1))$ belongs to A. Also, from inequality (3.8) we have that

$$< z, z^* > \leq k, \tag{3.10}$$

for all $z \in -P$. In particular as $0 \in -P$ we have $k \geq 0$. Suppose for some $z \in -P$, $< z, z^* > > 0$. Then we have $< \alpha z, z^* > = \alpha < z, z^* > \to \infty$ as $\alpha \to \infty$. However as P is a cone and $\alpha \geq 0$, $\alpha z \in -P$ if $z \in -P$. Therefore $< \alpha z, z^* > \leq k < \infty$ if $z \in -P$. Thus we have a contradiction and therefore

$$< z, z^* > \leq 0 \text{ for all } z \in -P \text{ and } k \geq 0. \tag{3.11}$$

As $-g(x_1) \in int(P)$ we have that there exists an $\epsilon > 0$ in R such that $||z||_Z \leq \epsilon$ implies that $-g(x_1) + z \in P$. Therefore, from (3.11) we have that $< g(x_1) - z, z^* > \leq 0$ if $||z||_Z \leq \epsilon$ which implies that $< g(x_1), z^* > \leq < z, z^* >$ if $||z||_Z \leq \epsilon$. From inequality (3.9) we have $0 \leq k \leq < g(x_1), z^* > \leq < z, z^* >$ if $||z||_Z \leq \epsilon$. This implies that for any $z \in Z$, $< z, z^* > \geq 0$. For any nonzero $z \in Z$,

$$\left|\left|\frac{\epsilon z}{||z||_Z}\right|\right|_Z \leq \epsilon$$

and therefore $< \frac{\epsilon z}{||z||_Z}, z^* > \geq 0$. This implies that for any $z \in Z$, $< z, z^* > \geq 0$. As Z is a vector space (which implies $- < z, z^* > \geq 0$) it follows that

$< z, z^* >= 0$ for all $z \in Z$. Thus $z^* = 0$. This contradicts $(z, s) \neq (0,0)$ and therefore, $s > 0$.

Let $z_0^* = \frac{z^*}{s}$. Dividing inequality (3.7) by s we have

$$< z, z_0^* > + r \geq \frac{k}{s} \text{ for all } (z, r) \in A \text{ and} \tag{3.12}$$

dividing inequality (3.8) by s we have

$$< z, z_0^* > + r \leq \frac{k}{s} \text{ for all } (z, r) \in B. \tag{3.13}$$

In particular, as $(z, \mu_0) \in B$ for all $z \in -P$ it follows from inequality (3.13) that

$$< z, z_0^* > \leq \frac{k}{s} - \mu_0 \text{ for all } z \in -P.$$

This implies that $< z, z_0^* > \leq 0$ for all $z \in -P$. Indeed, if for some $z_1 \in -P, < z_1, z_0^* > > 0$ then $< \alpha z_1, z^* > \to \infty$ as $\alpha \to \infty$ which contradicts the fact that $< \alpha z_1, z^* >$ is bounded above by $\frac{k}{s} - \mu_0$. Thus we conclude that $z_0^* \in P^{\oplus}$.

Also, as $(g(x), f(x))$ for $x \in \Omega$ is in A it follows from (3.12) that

$$< g(x), z_0^* > + f(x) \geq \frac{k}{s} \text{ for all } x \in \Omega \text{ and} \tag{3.14}$$

as $(0, \mu_0) \in B$ it folllows from (3.13) that

$$\mu_0 \leq \frac{k}{s} \text{ for all } (z, r) \in B. \tag{3.15}$$

From inequalities (3.14) and (3.15) we conclude that

$$\inf\{< g(x), z_0^* > + f(x) : x \in \Omega\} \geq \mu_0. \tag{3.16}$$

Suppose $x \in \Omega$ and $g(x) \leq 0$ (i.e. x is feasible), then

$$f(x) + < g(x), z_0^* > \leq f(x), \tag{3.17}$$

because $z_0^* \in P^{\oplus}$. Therefore, we have

$$\begin{aligned} \inf\{f(x) + < g(x), z_0^* >: x \in \Omega\} &\leq \inf\{f(x) + < g(x), z_0^* > \\ &\qquad : x \in \Omega, g(x) \leq 0\} \\ &\leq \inf\{f(x) : x \in \Omega, g(x) \leq 0\} = \mu_0. \end{aligned}$$

The first inequality is true because $\Omega \supset \{x \in \Omega, g(x) \leq 0\}$ and the second inequality follows from (3.17).

It follows from inequality (3.16) that

$$\mu_0 = \inf\{f(x) + < g(x), z_0^* >: x \in \Omega\}. \tag{3.18}$$

Let x_0 be such that $x_0 \in \Omega$ and $g(x_0) \leq 0$ and $f(x_0) = \mu_0$. Then

$$f(x_0) = \mu_0 \leq f(x_0) + < g(x_0), z_0^* > \leq f(x_0) = \mu_0.$$

The first inequality follows from equation (3.18) and the second inequality is true because $z_0^* \in P^{\oplus}$ and $g(x_0) \leq 0$. This proves that $< g(x_0), z_0^* >= 0$. □

Lemma 3.3.8. *Let X be a Banach space, Ω be a convex subset of X, Y be a finite dimensional normed space, Z be a normed space with positive cone P. Let Z^* denote the dual space of Z with a postive cone $P^{\oplus}$. Let $f : \Omega \to R$ be a real valued convex functional, $g : X \to Z$ be a convex mapping, $H : X \to Y$ be an affine linear map and $0 \in int(\{y \in Y : H(x) = y$ for some $x \in \Omega\})$. Define*

$$\mu_0 := \inf\{f(x) : g(x) \leq 0, \ H(x) = 0, \ x \in \Omega\}. \tag{3.19}$$

Suppose there exists $x_1 \in \Omega$ such that $g(x_1) < 0$ and $H(x_1) = 0$ and suppose μ_0 is finite. Then, there exist $z_0^* \geq 0$ and y_0^* such that

$$\mu_0 = \inf\{f(x) + < g(x), z_0^* > + < H(x), y_0^* >: x \in \Omega\}. \tag{3.20}$$

Proof. Let

$$\Omega_1 := \{x : \ x \in \Omega, \ H(x) = 0\}.$$

Applying Lemma 3.3.7 to Ω_1 we know that there exists $z_0^* \in P^{\oplus}$ such that

$$\mu_0 = \inf\{f(x) + < g(x), z_0^* >: x \in \Omega_1\}. \tag{3.21}$$

Consider the convex subset,

$$H(\Omega) := \{y \in Y : \ y = H(x) \text{ for some } x \in \Omega\}$$

of Y. For $y \in H(\Omega)$ define

$$k(y) := \inf\{f(x) + < g(x), z_0^* >: \ x \in \Omega, \ H(x) = y\}.$$

We now show that k is convex. Suppose $y, y' \in H(\Omega)$ and x, x' are such that $H(x) = y$ and $H(x') = y'$. Suppose, $0 < \lambda < 1$. We have, $\lambda(f(x) + < g(x), z_0^* >) + (1-\lambda)(f(x') + < g(x'), z_0^* >) \geq f(\lambda x + (1-\lambda)x') + < g(\lambda x + (1-\lambda)x'), z_0^* > \geq k(\lambda y + (1-\lambda)y')$. (the first inequality follows from the convexity of f and g. The second inequality is true because $H(\lambda x + (1-\lambda)x') = \lambda y + (1-\lambda)y'$.) Taking infimum on the left hand side we obtain $\lambda k(y) + (1-\lambda)k(y') \geq k(\lambda y + (1-\lambda)y')$. This proves that k is a convex function.

We now show that $k : H(\Omega) \to R$ (i.e. we show that $k(y) > -\infty$ for all $y \in H(\Omega)$). As, $0 \in int[H(\Omega)]$ we know that there exists an $\epsilon > 0$ such that if $||y|| \leq \epsilon$ then $y \in H(\Omega)$. Take any $y \in H(\Omega)$ such that $y \neq 0$. Choose λ, y' such that

$$\lambda = \frac{\epsilon}{2||y||} \text{ and } y' = -\lambda y.$$

This implies that $y' \in H(\Omega)$. Let, $\beta = \frac{\lambda}{\lambda + 1}$. We have

$$(1-\beta)y' + \beta y = 0.$$

Therefore, from convexity of the function k we have

$$\beta k(y) + (1-\beta)k(y') \geq k(0) = \mu_0.$$

Note that $\mu_0 > -\infty$ by assumption. Therefore, $k(y) > -\infty$. Note, that for all $y \in H(\Omega)$, $k(y) < \infty$. This proves that k is a real valued function.

Let $[k, H(\Omega)]$ be defined as given below

$$[k, H(\Omega)] := \{(r, y) \in R \times Y : y \in H(\Omega),\ k(y) \leq r\}.$$

We first show that $[k, H(\Omega)]$ has nonempty interior. As, k is a real valued convex function on the finite-dimensional convex set $H[\Omega]$ and $0 \in int[H(\Omega)]$ we have from from Lemma 3.1.2 that k is continuous at 0. Let $r_0 = k(0) + 2$ and choose ϵ' such that $0 < \epsilon' < 1$. As, k is continuous at 0 we know that there exists $\delta > 0$ such that $y \in H(\Omega)$ and $||y|| \leq \delta$ implies that

$$|k(y) - k(0)| < \epsilon'.$$

This means that if $y \in H(\Omega)$ and $||y|| \leq \delta$ then

$$k(y) < k(0) + \epsilon' < k(0) + 1 < r_0 - \frac{1}{2}.$$

Therefore, for all $y \in H(\Omega)$ with $||y|| \leq \delta$ we have $k(y) < r_0 - \frac{1}{2}$. This implies that for all $(r, y) \in R \times Y$ such that $|r - r_0| < \frac{1}{4}, y \in H(\Omega)$ and $||y|| \leq \delta$ we have $k(y) < r$. This proves that $(r_0, 0) \in int([k, H(\Omega)])$.

It is clear that $(k(0), 0) \in R \times Y$ is not in the interior of $[k, H(\Omega)]$. Using, Corollary 3.2.2 we know that there exists $(s, y^*) \neq (0, 0) \in R \times Y^*$ such that for all $(r, y) \in [k, H(\Omega)]$ the following is true

$$< y, y^* > + rs \geq < 0, y^* > + k(0)s = s\mu_0. \tag{3.22}$$

In particular, $rs \geq s\mu_0$ for all $r \geq \mu_0$ (note that $(r, 0) \in [k, H(\Omega)]$ for all $r \geq \mu_0$). This means that $s \geq 0$.

Suppose, $s = 0$. We have from (3.22) that $< y, y^* > \geq 0$ for all $y \in H(\Omega)$. As, $0 \in int[H(\Omega)]$ it follows that there exists an $\epsilon \in R$ such that $||y|| \leq \epsilon$ implies that $< y, y^* > \geq 0$ and $< -y, y^* > \geq 0$. This implies that if $||y|| \leq \epsilon$ then $< y, y^* > = 0$. But, then for any $y \in Y$ one can choose a positive constant α such that $||\alpha y|| \leq \epsilon$ and therefore $< \alpha y, y^* > = 0$. This implies that $(s, y^*) = (0, 0)$ which is not possible. Therefore, we conclude that $s > 0$.

Let $y_0^* = y^*/s$. From (3.22) we have,

$$< y, y_0^* > + r \geq \mu_0, \quad \text{for all } (r, y) \in [k, H(\Omega)]. \tag{3.23}$$

This implies that for all $y \in H(\Omega)$,

$$< y, y_0^* > + k(y) \geq \mu_0, \tag{3.24}$$

(This is because $(k(y), y) \in [k, H(\Omega)]$). Therefore, for all $x \in \Omega$,

$$< H(x), y_0^* > + f(x) + < g(x), z_0^* > \geq \mu_0, \tag{3.25}$$

which implies that

$$\inf\{f(x)+ < g(x), z_0^* > + < H(x), y_0^* >: \ x \in \Omega\} \geq \mu_0. \tag{3.26}$$

But if $x \in \Omega$ is such that $H(x) = 0$ then

$$\begin{aligned} f(x)+ < g(x), z_0^* > &= f(x)+ < g(x), z_0^* > + < H(x), y_0^* > \\ &\geq \inf\{f(x)+ < g(x), z_0^* > + < H(x), y_0^* >: \ x \in \Omega\} \\ &\geq \mu_0. \end{aligned}$$

Taking infimum on the left hand side of the above inequality over all $x \in \Omega$ which satisfy $H(x) = 0$ (that is infimum over all $x \in \Omega_1$) we have,

$$\mu_0 = \inf\{f(x)+ < g(x), z_0^* > + < H(x), y_0^* >: \ x \in \Omega\}. \tag{3.27}$$

This proves the lemma. □

The following is a Lagrange duality theorem.

Theorem 3.3.2 (Kuhn-Tucker-Lagrange duality). *Let X be a Banach space, Ω be a convex subset of X, Y be a finite dimensional normed space, Z be a normed space with positive cone P. Let Z^* denote the dual space of Z with a positive cone $P^\oplus$. Let $f : \Omega \to \mathbb{R}$ be a real valued convex functional, $g : X \to Z$ be a convex mapping, $H : X \to Y$ be an affine linear map and $0 \in int[range(H)]$. Define*

$$\mu_0 := \inf\{f(x) : \ g(x) \leq 0, \ H(x) = 0, \ x \in \Omega\}. \tag{3.28}$$

Suppose there exists $x_1 \in \Omega$ such that $g(x_1) < 0$ and $H(x_1) = 0$ and suppose μ_0 is finite. Then,

$$\mu_0 = \max\{\varphi(z^*, y) : z^* \geq 0, \ z^* \in Z^*, \ y \in Y\}, \tag{3.29}$$

where $\varphi(z^*, \ y) := \inf\{f(x)+ < g(x), z^* > + < H(x), y >: x \in \Omega \}$ and the maximum is achieved for some $z_0^* \geq 0$, $z_0^* \in Z^*$, $y_0 \in Y$.

Furthermore if the infimum in (3.28) is achieved by some $x_0 \in \Omega$ then

$$< g(x_0), z_0^* > + < H(x_0), y_0 >= 0, \tag{3.30}$$

and x_0 minimizes

$$f(x)+ < g(x), z_0^* > + < H(x), y_0 >, \quad \text{over all } x \in \Omega. \tag{3.31}$$

Proof. Given any $z^* \geq 0$, $y \in Y$ we have

$$\begin{aligned} \inf_{x\in\Omega}\{f(x)+ < g(x), z^* > \\ + < H(x), y >\} &\leq \inf_{x\in\Omega}\{f(x)+ < g(x), z^* > + < H(x), y > \\ &\qquad : \ g(x) \leq 0, \ H(x) = 0\} \\ &\leq \inf_{x\in\Omega}\{f(x) : \ g(x) \leq 0, \ H(x) = 0\} \\ &= \mu_0. \end{aligned}$$

Therefore it follows that $\max\{\varphi(z^*, y) : z^* \geq 0, \ y \in Y\} \leq \mu_0$. From Lemma 3.3.8 we know that there exists $z_0^* \in Z^*, z_0^* \geq 0, \ y_0 \in Y$ such that $\mu_0 = \varphi(z_0^*, \ y_0)$. This proves (3.29).

Suppose there exists $x_0 \in \Omega, H(x_0) = 0,\ g(x_0) \le 0$ and $\mu_0 = f(x_0)$ then $\mu_0 = \varphi(z_0^*,\ y_0) \le f(x_0) + < g(x_0), z_0^* > + < H(x_0), y_0 > \le f(x_0) = \mu_0$. Therefore we have $< g(x_0), z_0^* > + < H(x_0), y_0 >= 0$ and $\mu_0 = f(x_0) + < g(x_0), z_0^* > + < H(x_0), y_0 >$. This proves the theorem. □

We refer to (3.28) as the *Primal* problem and (3.29) as the *Dual* problem.

Corollary 3.3.1 (Sensitivity). *Let X, Y, Z, f, H, g, Ω be as in Theorem 3.3.2. Let x_0 be the solution to the problem*

$$\text{minimize } f(x)$$
$$\text{subject to } x \in \Omega,\ H(x) = 0,\ g(x) \le z_0$$

with (z_0^*, y_0) as the dual solution. Let x_1 be the solution to the problem

$$\text{minimize } f(x)$$
$$\text{subject to } x \in \Omega,\ H(x) = 0,\ g(x) \le z_1$$

with (z_1^*, y_1) as the dual solution. Then,

$$< z_1 - z_0, z_1^* > \le f(x_0) - f(x_1) \ \le < z_1 - z_0, z_0^* > . \tag{3.32}$$

Proof. From Theorem 3.3.2 we know that for any $x \in \Omega$,

$$\begin{aligned} & f(x_0) + < g(x_0) - z_0, z_0^* > + < H(x_0), y_0 > \\ \le\ & f(x) + < g(x) - z_0, z_0^* > + < H(x), y_0 > . \end{aligned}$$

In particular we have

$$\begin{aligned} & f(x_0) + < g(x_0) - z_0, z_0^* > + < H(x_0), y_0 > \\ \le\ & f(x_1) + < g(x_1) - z_0, z_0^* > + < H(x_1), y_0 > . \end{aligned}$$

From Theorem 3.3.2 we know that $< g(x_0) - z_0, z_0^* > + < H(x_0), y_0 >= 0$ and $H(x_1) = 0$. This implies

$$f(x_0) - f(x_1) \ \le < g(x_1) - z_0, z_0^* > \le < z_1 - z_0, z_0^* > .$$

A similar argument gives the other inequality. This proves the corollary. □

4. Paradigm for Control Design

We present notions of stability, causality and well-posedness of interconnections of systems. The main part of this chapter focusses on the parametrization of all closed loop maps that are achievable through stabilizing controllers.

4.1 Notation and Preliminaries

We will generalize the ℓ_p space that we introduced in Chapter 2. Let ℓ^n denote the space of all vector-valued real sequences taking values on positive integers that is

$$\ell^n = \{x : x = (x_1, x_2, \ldots, x_n) \text{ with } x_i \in \ell_p\}.$$

For any x in ℓ^n let

$$\begin{aligned} \|x\|_p &= \left(\sum_{k=0}^{\infty}\sum_{i=1}^{n} |x_i(k)|^p\right)^{\frac{1}{p}} \quad 1 \leq p < \infty \text{ and} \\ \|x\|_\infty &= \sup_k \max_i |x_i(k)| \end{aligned}$$

where $x = (x_1, x_2, \ldots, x_n)$ and $x_i = (x_i(0), x_i(1), \ldots)$ with $x_i(k) \in R$. Let

$$\ell_p^n := \{x | x \in \ell^n, \|x\|_p < \infty\}.$$

All the results that were established for ℓ_p spaces in Section 2.4 hold for the $(\ell_p^n, \|\cdot\|_p)$ spaces. We often refer to ℓ^n as a *signal-space*. Let $\ell_p^{m\times n}$ denote the spaces of $m \times n$ matrices with each element of the matrix in ℓ_p. Let P_k denote the truncation operator on $\ell^{m\times n}$ which is defined by

$$P_k(x(0), x(1), x(2), \ldots) = (x(0), x(1), \ldots, x(k), 0, 0, \ldots).$$

Let S denote the shift map from ℓ^n to ℓ^n defined by

$$S(x(0), x(1), x(2), \ldots) = (0, x(0), x(1), x(2), x(3), \ldots).$$

Definition 4.1.1 (Causality). *A linear map $T : \ell^n \to \ell^m$ is said to be causal if*

$$P_t T = P_t T P_t \text{ for all } t.$$

$\mathcal{T}$ is strictly causal if

$$P_t\mathcal{T} = P_t\mathcal{T}P_{t-1} \text{ for all } t$$

where P_t is the truncation operator.

Definition 4.1.2 (Time invariance). *A map $\mathcal{T} : \ell^n \to \ell^m$ is time invariant if $S\mathcal{T} = \mathcal{T}S$ where S is the shift operator.*

Let $\mathcal{T}$ be a linear map from $(\ell_p^n, ||.||_p)$ to $(\ell_p^m, ||.||_p)$. The p-induced norm of $\mathcal{T}$ is defined as

$$||\mathcal{T}||_{p-ind} := \sup_{||x||_p \neq 0} \frac{||\mathcal{T}x||_p}{||x||_p}.$$

We often refer to a map from a signal-space ℓ^n to another signal space ℓ^m as a *system*.

Definition 4.1.3 (Stability). *A linear map $\mathcal{T} : (X, ||.||_X) \to (Y, ||.||_Y)$ is said to be stable if it is bounded. $\mathcal{T} : (\ell_p^n, ||.||_p) \to (\ell_p^m ||.||_p)$ is said to be ℓ_p stable if it is bounded.*

Example 4.1.1. Let $\mathcal{T} : \ell \to \ell$ be a linear operator such that $y = \mathcal{T}u$ is defined as

$$\begin{pmatrix} y(0) \\ y(1) \\ y(2) \\ \vdots \end{pmatrix} = \begin{pmatrix} T(0) & 0 & \ldots & \\ T(1) & T(0) & 0 & \ldots \\ T(1) & T(2) & T(0) & \ldots \\ \vdots & \vdots & \ddots & \vdots \end{pmatrix} \begin{pmatrix} u(0) \\ u(1) \\ u(2) \\ \vdots \end{pmatrix},$$

where $y = (y(0),\ y(1),\ y(2),\ \ldots)$, $u = (u(1),\ u(2),\ \ldots)$ and $T(j) \in R$, for all $j = 0, 1, \ldots$. Let t be any positive integer. Then it can be verified that $P_t\mathcal{T}P_tu = P_t\mathcal{T}u$ for any $u \in \ell$. Thus $\mathcal{T}$ is causal. If $T(0) = 0$ then it can be verified that $P_t\mathcal{T}P_{t-1} = P_t\mathcal{T}$. In this case $\mathcal{T}$ is strictly causal.

It also follows that for any $u \in \ell$, $S\mathcal{T}u = \mathcal{T}Su$. Thus $\mathcal{T}$ is a time invaraint map.

Definition 4.1.4 (Convolution maps). *$\mathcal{T} : \ell^n \to \ell^m$ is linear, time invariant causal, convolution map if and only if $y = \mathcal{T}u$ is given by*

$$\begin{pmatrix} y_1 \\ y_2 \\ \vdots \\ y_m \end{pmatrix} = \begin{pmatrix} \mathcal{T}_{11} & \mathcal{T}_{12} & \ldots & \mathcal{T}_{1n} \\ \mathcal{T}_{21} & \mathcal{T}_{22} & \ldots & \mathcal{T}_{2n} \\ \vdots & \vdots & \vdots & \vdots \\ \mathcal{T}_{m1} & \mathcal{T}_{m2} & \ldots & \mathcal{T}_{mn} \end{pmatrix} \begin{pmatrix} u_1 \\ u_2 \\ \vdots \\ u_n \end{pmatrix},$$

where $y = (y_1\ y_2,\ \ldots,\ y_m) \in \ell^m$ and $u = (u_1,\ u_2,\ \ldots,\ u_n) \in \ell^n$, $\mathcal{T}_{ij} : \ell \to \ell$ is described by

$$\begin{pmatrix} y_i(0) \\ y_i(1) \\ y_i(2) \\ \vdots \end{pmatrix} = \begin{pmatrix} T_{ij}(0) & 0 & \dots & \\ T_{ij}(1) & T_{ij}(0) & 0 & \dots \\ T_{ij}(2) & T_{ij}(1) & T_{ij}(0) & \dots \\ \vdots & \vdots & \ddots & \vdots \end{pmatrix} \begin{pmatrix} u_j(0) \\ u_j(1) \\ u_j(2) \\ \vdots \end{pmatrix}, \tag{4.1}$$

with $T_{ij}(k) \in R$ for all $k = 0, 1, \dots$. $\{T_{ij}(k)\}_{k=0}^{\infty}$ is also called the impulse response of the system $\mathcal{T}_{ij}$. The linear map $\mathcal{T}_{ij}$ can be identified with the sequence $\{T_{ij}(k)\}_{k=0}^{\infty} \in \ell$. Thus with some abuse of notation we often write T_{ij} to mean the map $\mathcal{T}_{ij}$ and T to mean the map $\mathcal{T}$. Depending on the context T_{ij} can denote the map $\mathcal{T}_{ij}$ or the sequence $\{T_{ij}(k)\}$. The operation given by Equation 4.1 is often written as $y_i = T_{ij} * u_j$.

Lemma 4.1.1. *Let $T : \ell \to \ell$ be a linear, time invariant, causal, convolution map. Let $\{T(k)\}$ denote its impulse response. Then*

$$||T||_{\infty-ind} = \sum_{k=0}^{\infty} |T(k)|.$$

Proof. Note that for any $u \in \ell$ with $||u||_\infty \le 1$ we have

$$\begin{aligned} ||Tu||_\infty &= || \begin{pmatrix} T(0) & 0 & \dots & \\ T(1) & T(0) & 0 & \dots \\ T(2) & T(2) & T(0) & \dots \\ \vdots & \vdots & \ddots & \vdots \end{pmatrix} \begin{pmatrix} u(0) \\ u(1) \\ u(2) \\ \vdots \end{pmatrix} ||_\infty \\ &= \sup_n |\sum_{k=0}^{n} T(n-k)u(k)| \le \sup_n \sum_{k=0}^{n} |T(n-k)|\ |u(k)| \\ &\le \sup_n \sum_{k=0}^{\infty} |T(n-k)| = \sum_{k=0}^{\infty} |T(k)|. \end{aligned}$$

Thus we have shown that $||T||_{\infty-ind} \le \sum_{k=0}^{\infty} |T(k)|$.

Consider elements u^i in ℓ which are defined by

$$\begin{aligned} u_i(k) &= sgn(T(k)) \text{ if } k \le i \\ &= 0 \text{ if } k > i. \end{aligned}$$

It can be seen that if $y^i = Tu^i$ then $y^i(i) = \sum_{k=0}^{i} |T(k)|$. Thus it follows that $Tu^i \to \sum_{k=0}^{\infty} |T(k)|$. Note that $u^i \in \ell_\infty$ for all i. Thus it follows that $||T||_\infty \ge \sum_{k=0}^{\infty} |T(k)|$. This proves the lemma. □

From the lemma above it follows that for a linear, time invariant, causal, convolution map, $T : \ell^n \to \ell^m$ if $\sum_{t=0}^{\infty} |T_{ij}(t)| < \infty$ for all i and j, then T is a bounded map from ℓ_∞^n to ℓ_∞^m.

Lemma 4.1.2. *Let T be a linear, time invariant, causal map from $(\ell_\infty^n, ||\cdot||_\infty)$ to $(\ell_\infty^m, ||\cdot||_\infty)$. Then*

$$||T||_{\infty-ind} := \max_{1 \le i \le m} \sum_{j=1}^{n} ||T_{ij}||_1,$$

where T_{ij} are elements of T (see Definition 4.1.4) and $||T_{ij}||_1 := \sum_{k=0}^{\infty} |T_{ij}(k)|$.

Proof. The proof is left to the reader. □

Definition 4.1.5 (λ-transforms). *For a linear, time invariant, causal, convolution map $T : \ell^n \to \ell^m$ as described in Definition 4.1.4 the λ-transform of T is defined as*

$$\hat{T} := \sum_{i=0}^{\infty} T(i)\lambda^i,$$

where

$$T(i) := \begin{pmatrix} T_{11}(i) & T_{12}(i) & \dots & T_{1n}(i) \\ T_{21}(i) & T_{22}(i) & \dots & T_{2n}(i) \\ \vdots & \vdots & \vdots & \vdots \\ T_{m1}(i) & T_{m2}(i) & \dots & T_{mn}(i) \end{pmatrix}.$$

It can be shown that $\hat{T}$ is analytic inside the open unit disc and continuous on the boundary if the matrix sequence $\{T(i)\} \in \ell_1^{m\times n}$.

Example 4.1.2. Consider a linear map as described in Example 4.1.1 where $T(k) = \frac{1}{2^k}$. Then

$$\hat{T}(\lambda) = \sum_{k=0}^{\infty} T(k)\lambda^k = \sum_{k=0}^{\infty} \frac{1}{2^k}\lambda^k = \frac{1}{1-\frac{1}{2}\lambda}.$$

The ℓ_∞ induced norm of T, given by $||T||_{\infty-ind} = \sum_{k=0}^{\infty} |T(k)| = 2 < \infty$. Thus T is ℓ_∞ stable.

Definition 4.1.6 (Finite dimensional system). *If the λ-transform of any linear, time invariant, causal, convolution map $T : \ell^n \to \ell^m$ is such that $\hat{T}_{ij}(\lambda)$ is a ratio of finite polynomials in λ then T represents a finite dimensional system.*

We use the term FDLTIC as an abbreviation for finite-dimensional, linear, time invariant, causal. Unless otherwise mentioned all systems will be assumed to be convolution maps.

Lemma 4.1.3. *Let T be a FDLTIC system; $T : \ell^n \to \ell^m$. Then there exist real matrices A, B, C and D such that if $y = Tu$ for some $u \in \ell^n$ then*

$$\begin{aligned} x(k+1) &= Ax(k) + Bu(k) \\ y(k) &= Cx(k) + Du(k) \\ x(0) &= 0, \end{aligned} \tag{4.2}$$

where $u = (u(0), u(1), \dots)$ and $y = (y(0), y(1), \dots)$.

Proof. See [2]. □

The representation of the map T as given in (4.2) is called a *state space* representation of T. A convenient notation empployed to denote the system described by (4.2) is $\left[\begin{array}{c|c} A & B \\ \hline C & D \end{array}\right]$.

Definition 4.1.7 (Controllability, observability). *The pair of real matrices A and B with $A \in R^{n\times n}$ and with $B \in R^{n\times m}$ is a controllable pair if the matrix $[B,\ AB, \ldots, A^{n-1}B]$ has full rank. The pair of real matrices A and C is observable if (A^T, C^T) is controllable.*

Definition 4.1.8 (Stabilzability, detectability). *The pair of real matrices A and B with $A \in R^{n\times n}$ and with $B \in R^{n\times m}$ is a stabilizable pair if there exists a real matrix K such that $\rho(A + BK) < 1$, where ρ denotes the spectral radius. The pair of real matrices A and C is detectable if there exists a real matrix L such that $\rho(A + LC) < 1$.*

Definition 4.1.9 (Minimal realization). *The triplet (A, B, C) are a minimal realization of $\hat{T}$ in (4.2) if (A, B) is controllable and (A, C) is observable.*

Lemma 4.1.4. *Suppose T is a FDLTIC system. Then there exists a state-space description $\left[\begin{array}{c|c} A & B \\ \hline C & D \end{array}\right]$ of T such that (A, B, C) is minimal.*

Proof. See [2].

Definition 4.1.10 (Unimodular matrices). *A square polynomial matrix function $\hat{P}(\lambda) = P(0) + P(1)\lambda + \ldots + P(k)\lambda^k$, is said to be unimodular if the determinant of $\hat{P}(\lambda)$ is a non-zero constant independent of λ.*

Theorem 4.1.1. *Let $\hat{T}(\lambda)$ be a $m \times n$ matrix of rational functions of λ (a function is a rational function of λ if it can be written as a ratio of two polynomials of λ). Then there exist $\hat{L}$, $\hat{U}$ and $\hat{M}$ such that $\hat{T} = \hat{L}\hat{M}\hat{U}$ where $\hat{L}$ and $\hat{U}$ are unimodular with appropriate dimensions and $\hat{M}$ has the structure*

$$\hat{M} = \begin{pmatrix} \frac{\hat{\epsilon}_1}{\hat{\psi}_1} & & & 0 \ldots 0 \\ & \ddots & & \vdots \ \ddots \ \vdots \\ & & \frac{\hat{\epsilon}_r}{\hat{\psi}_r} & 0 \ldots 0 \\ 0 & \ldots & 0 & 0 \ldots 0 \\ \vdots & \ddots & \vdots & 0 \ldots 0 \\ 0 & \ldots & 0 & 0 \ldots 0 \end{pmatrix}.$$

$\{\hat{\epsilon}_i, \hat{\psi}_i\}$ are coprime (that is they do not have any common factors) monic (leading coefficient is one) polynomials, which are not identically zero for all $i = 1, \ldots, r$ with the following divisibility property: $\hat{\epsilon}_i(\lambda)$ divides $\hat{\epsilon}_{i+1}(\lambda)$ without remainder and $\hat{\psi}_{i+1}(\lambda)$ divides $\hat{\psi}_i(\lambda)$ without remainder.

$\hat{M}$ is called the *Smith-Mcmillan* form of $\hat{T}(\lambda)$.

Definition 4.1.11 (Zeros and poles of $\hat{T}$). *The zeros of $\hat{T}(\lambda)$ are the roots of $\Pi_{i=1}^{r}\hat{\epsilon}_i(\lambda)$. The poles of $\hat{T}(\lambda)$ are the roots of $\Pi_{i=1}^{r}\hat{\psi}_i(\lambda)$.*

Theorem 4.1.2. *Suppose T is a FDLTIC system. Then the following statements are equivalent.*

1. *T is ℓ_p stable for any p, $1 \leq p \leq \infty$.*
2. *If $\left[\begin{array}{c|c} A & B \\ \hline C & D \end{array}\right]$ is any state-space description of T such that (A, B) is stabilizable and (A, C) is detectable then $\rho(A) < 1$ where $\rho(A)$ denotes the spectral radius of the matrix A.*
3. *$\hat{T}(\lambda)$ the λ-transform of T has all its poles outside the unit disc (that is if λ_0 is a pole of $\hat{T}$ then $|\lambda_0| > 1$).*

Proof. See [2]. □

Example 4.1.3. Consider Example 4.1.2. T is a FDLTIC system. A state-space description of T is given by $\left[\begin{array}{c|c} A & B \\ \hline C & D \end{array}\right] = \left[\begin{array}{c|c} 0.5 & 0.5 \\ \hline 1 & 1 \end{array}\right]$. Note that as T is ℓ_∞ stable it is ℓ_p stable for all $1 \leq p \leq \infty$. Furthermore $\hat{T}(\lambda)$ has a single pole at 2 which is outside the unit disc.

This theorem establishes the fact that for FDLTIC systems stability in ℓ_p sense implies stability in ℓ_q sense for any p and q such that $1 \leq p \leq \infty$ and $1 \leq q \leq \infty$. Thus for FDLTIC systems we can use the term stability to mean stability in ℓ_p sense for any $1 \leq p \leq \infty$.

Definition 4.1.12 (Normal rank). *Let $\hat{T}(\lambda)$ be a $m \times n$ matrix of rational functions of λ. The normal rank of $\hat{T}$ is the rank of $\hat{T}(\lambda_0)$ where λ_0 is any complex number which is not a zero of $\hat{T}(\lambda)$.*

Definition 4.1.13 (rcf, lcf, dcf). *Stable FDLTIC systems M and N are right coprime if there exist stable FDLTIC systems X and Y such that the λ-transforms satisfy the identity*

$$\hat{X}(\lambda)\hat{M}(\lambda) - \hat{Y}(\lambda)\hat{N}(\lambda) = I. \tag{4.3}$$

Stable FDLTIC systems M and N are left coprime if there exist stable FDLTIC systems X and Y such that the λ-transforms satisfy the identity

$$\hat{M}(\lambda)\hat{X}(\lambda) - \hat{N}(\lambda)\hat{Y}(\lambda) = I. \tag{4.4}$$

Suppose $\hat{T} = \hat{N}\hat{M}^{-1} = \hat{\tilde{M}}^{-1}\hat{\tilde{N}}$ where N and M are right coprime and $\tilde{M}$ and $\tilde{N}$ are left coprime. Then the pair N and M form a right coprime factorization (rcf) of T and the pair $\tilde{M}$ and $\tilde{N}$ form a left coprime factorization (lcf) of T.

A doubly-coprime factorization (dcf) of a FDLTIC system T is a set of stable FDLTIC maps $M, N, \tilde{M}$ and $\tilde{N}$ such that $\hat{T} = \hat{N}\hat{M}^{-1} = \hat{\tilde{M}}^{-1}\hat{\tilde{N}}$ and

$$\begin{pmatrix} \hat{\tilde{X}} & -\hat{\tilde{Y}} \\ -\hat{\tilde{N}} & \hat{\tilde{M}} \end{pmatrix} \begin{pmatrix} \hat{M} & \hat{Y} \\ \hat{N} & \hat{X} \end{pmatrix} = I. \tag{4.5}$$

Lemma 4.1.5. *Let T be a FDLTIC map with a state space description $\left[\begin{array}{c|c} A & B \\ \hline C & D \end{array}\right]$. Suoppse$(A, B)$ is stabilizable and (A, C) is detectable. Then there exists a dcf of T.*

Proof. In the definition of dcf let

$$X = \left[\begin{array}{c|c} A+BF & -L \\ \hline C+DF & I \end{array}\right], \quad Y = \left[\begin{array}{c|c} A+BF & -L \\ \hline F & 0 \end{array}\right],$$
$$\tilde{X} = \left[\begin{array}{c|c} A+LC & B+LD \\ \hline F & -I \end{array}\right], \quad \tilde{Y} = \left[\begin{array}{c|c} A+LC & L \\ \hline F & 0 \end{array}\right],$$
$$M = \left[\begin{array}{c|c} A+BF & -B \\ \hline F & -I \end{array}\right], \quad N = \left[\begin{array}{c|c} A+BF & -B \\ \hline C+DF & -D \end{array}\right],$$
$$\tilde{M} = \left[\begin{array}{c|c} A+LC & -L \\ \hline -C & I \end{array}\right], \text{ and } \tilde{N} = \left[\begin{array}{c|c} A+LC & B+LD \\ \hline C & D \end{array}\right].$$

Then it can be shown that $\hat{T} = \hat{N}\hat{M}^{-1} = \hat{\tilde{M}}^{-1}\hat{\tilde{N}}$ and (4.5) is satisfied. □

Definition 4.1.14 (Unit in ℓ_1). *A system U in $\ell_1^{m \times m}$ is a unit in ℓ_1 if U^{-1} is in $\ell_1^{m \times m}$.*

4.2 Interconnection of Systems

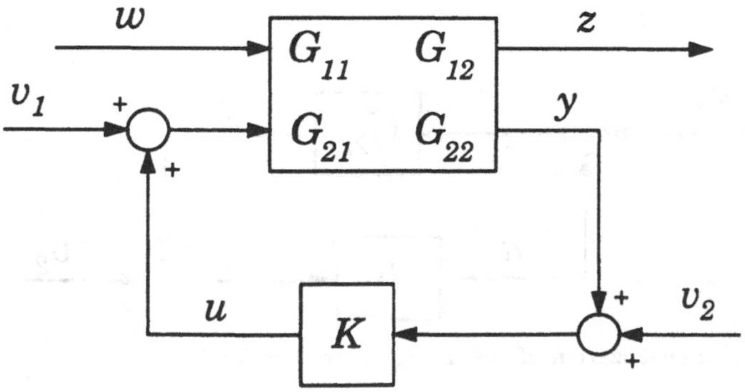

Fig. 4.1. Framework

Many control design issues can be cast into the framework shown in Figure 4.1 which shows a gerealized plant G, exogenous inputs w, control inputs u, measured outputs y and regulated outputs z. K is the controller which maps the measured outputs y to control inputs u when v_1 and v_2 are zero. Both K and G are linear time invariant causal convolution maps. With respect to the interconnection of systems G and K in Figure 4.1, the first issue that needs to addressed is the existence and uiqueness of signals z, u and y for given input signals w, v_1 and v_2. If the signals resulting in an interconnection of two systems exist (that is they satisfy the conditions posed by the interconnection) and are unique then we say that the interconnection is *well-posed.* Note that for the interconnection in Figure 4.1 the existence and uniqueness of z, u and y is sufficient for the well-posedness of the interconnection. The signals satisfy the relation

$$\begin{pmatrix} I & -G_{12} & 0 \\ 0 & I & -K \\ 0 & -G_{22} & I \end{pmatrix} \begin{pmatrix} z \\ u \\ y \end{pmatrix} = \begin{pmatrix} G_{11} & 0 & 0 \\ 0 & I & K \\ G_{21} & 0 & 0 \end{pmatrix} \begin{pmatrix} w \\ v_1 \\ v_2 \end{pmatrix}. \tag{4.6}$$

We will suppose throughout that the interconnection is well-posed. This is guaranteed if the map G_{22} is strictly causal. Let $H(G, K)$ be such that

$$\begin{pmatrix} z \\ u \\ y \end{pmatrix} = H(G, K) \begin{pmatrix} w \\ v_1 \\ v_2 \end{pmatrix}.$$

The interconnection described by $H(G, K)$ is often referred to as the *closed loop map.*

Definition 4.2.1 (Stability of closed loop maps). *The closed loop map described by Figure 4.1 is ℓ_p stable if $\|H(G, K)\|_{p-ind} < \infty$. In such a case K is said to be a stabilizing controller in the ℓ_p sense.*

4.2.1 Interconnection of FDLTIC Systems

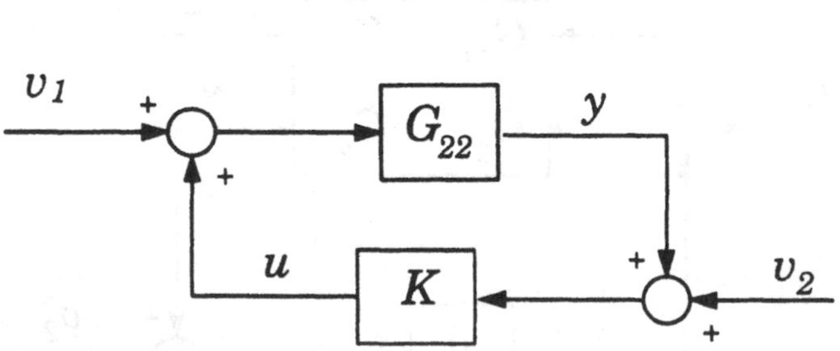

Fig. 4.2. Parametrization of stabilizing controllers for G_{22}.

Lemma 4.2.1. *Let* G_{22} *be a FDLTIC system which has a dcf given by* $\hat{G}_{22} = \hat{N}\hat{M}^{-1} = \tilde{\hat{M}}^{-1}\tilde{\hat{N}}$ *where*

$$\begin{pmatrix} \tilde{\hat{X}} & -\tilde{\hat{Y}} \\ -\tilde{\hat{N}} & \tilde{\hat{M}} \end{pmatrix} \begin{pmatrix} \hat{M} & \hat{Y} \\ \hat{N} & \hat{X} \end{pmatrix} = I. \tag{4.7}$$

A FDLTIC controller K *stabilizes the closed loop map shown in Figure 4.2 if and only if* K *has a rcf* $\hat{K} = \hat{Y}_1\hat{X}_1^{-1}$ *such that the map*

$$\begin{pmatrix} M & Y_1 \\ N & X_1 \end{pmatrix}$$

is a unit in ℓ_1.

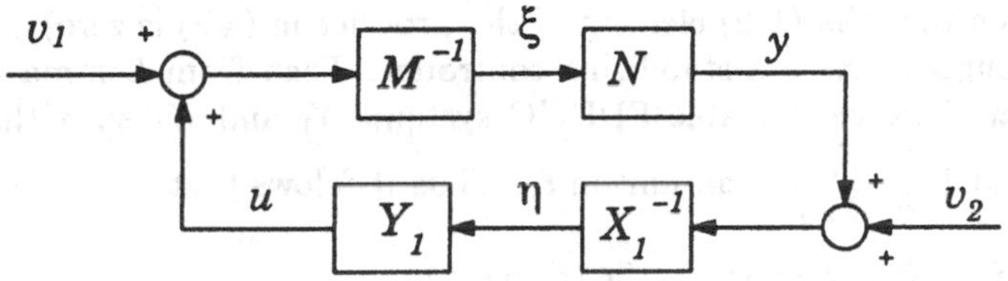

Fig. 4.3. Closed loop map with coprime factors for G_{22} and K.

Proof. ($\Leftarrow$) Suppose an rcf of K is given by $\hat{Y}_1\hat{X}_1^{-1}$ and suppose $\begin{pmatrix} M & Y_1 \\ N & X_1 \end{pmatrix}$ is an unit in ℓ_1. It is clear that the Figure 4.2 is the same as Figure 4.3. Note that the map from $(\xi,\ \eta)$ to $(v_1,\ v_2)$ is given by $\begin{pmatrix} M & -Y_1 \\ -N & X_1 \end{pmatrix}$. Because the inverse of this map is stable it follows that the map from $(v_1,\ v_2)$ to $(\xi,\ \eta)$ is stable. But $||y||_p = ||N\xi||_p \leq ||N||_{p-ind}||\xi||_p$ and $||u||_p = ||Y_1\eta||_p \leq ||Y_1||_{p-ind}||\eta||_p$. Thus the map from $(v_1,\ v_2)$ to $(u,\ y)$ is stable and therefore the closed loop map is stable.

($\Rightarrow$)Let FDLTIC controller K be such that the closed loop map in Figure 4.2 is stable. Thus the map from $(v_1,\ v_2)$ to $(u,\ y)$ is stable. Every FDLTIC system admits a dcf (see Lemma 4.1.5), and therefore it admits a rcf also. Let a rcf of K be given by $\hat{K} = \hat{Y}_1\hat{X}_1^{-1}$. From the dcf of G_{22} it is follows that $\tilde{\hat{X}}\hat{M} - \tilde{\hat{Y}}\hat{N} = I$. Multiplying both sides of this equation by ξ we have $\xi = \tilde{X}(v_1+u) - \tilde{Y}y$ and thus $||\xi||_p \leq ||\tilde{X}||_{p-ind}(||v_1||_p + ||u||_p) + ||\tilde{Y}||_{p-ind}||y||_p$. This implies that the map from $(v_1,\ v_2)$ to (ξ, η) is stable. Thus the inverse of the map $\begin{pmatrix} M & -Y_1 \\ -N & X_1 \end{pmatrix}$ is stable. □

Theorem 4.2.1. *Let FDLTIC system* G_{22} *admit a dcf as given in Lemma 4.2.1. Then* K *is a FDLTIC stabilizing controller for the closed loop system in Figure 4.2 if and only if*

$$\hat{K} = (\hat{Y} - \hat{M}\hat{Q})(\hat{X} - \hat{N}\hat{Q})^{-1} = (\hat{\tilde{X}} - \hat{Q}\hat{\tilde{N}})^{-1}(\hat{\tilde{Y}} - \hat{Q}\hat{\tilde{M}}),$$

for some FDLTIC stable system Q.

Proof. ($\Leftarrow$) Multiplying both sides of (4.7) by $\begin{pmatrix} I & \hat{Q} \\ 0 & I \end{pmatrix}$ from the left and by $\begin{pmatrix} I & -\hat{Q} \\ 0 & I \end{pmatrix}$ from the right we have

$$\begin{pmatrix} \hat{\tilde{X}} - \hat{Q}\hat{\tilde{N}} & -\hat{\tilde{Y}} + \hat{Q}\hat{\tilde{M}} \\ -\hat{\tilde{N}} & -\hat{\tilde{M}} \end{pmatrix} \begin{pmatrix} \hat{M} & \hat{Y} - \hat{M}\hat{Q} \\ \hat{N} & \hat{X} - \hat{N}\hat{Q} \end{pmatrix} = I, \tag{4.8}$$

where Q is a stable FDLTIC map. From Lemma 4.2.1 it follows that $\hat{K} = (\hat{Y} - \hat{M}\hat{Q})(\hat{X} - \hat{N}\hat{Q})^{-1}$ is a stabilizing controller. It also follows from (4.8) that $(\hat{Y} - \hat{M}\hat{Q})(\hat{X} - \hat{N}\hat{Q})^{-1} = (\hat{\tilde{X}} - \hat{Q}\hat{\tilde{N}})^{-1}(\hat{\tilde{Y}} - \hat{Q}\hat{\tilde{M}})$ (follows from the observation that the $(1,2)$ element of the product in (4.8) is zero).

($\Rightarrow$) Suppose K is a stabilizing controller. Then from Lemma 4.2.1 we know that there exist stable FDLTIC systems Y_1 and X_1 such that $K = Y_1 X_1^{-1}$ and $\begin{pmatrix} M & Y_1 \\ N & X_1 \end{pmatrix}$ is an unit in ℓ_1. Thus it folows that

$$\begin{pmatrix} \hat{\tilde{X}} & -\hat{\tilde{Y}} \\ -\hat{\tilde{N}} & \hat{\tilde{M}} \end{pmatrix} \begin{pmatrix} \hat{M} & \hat{Y}_1 \\ \hat{N} & \hat{X}_1 \end{pmatrix} = \begin{pmatrix} I & -\hat{Q}\hat{D} \\ 0 & \hat{D} \end{pmatrix}$$

is a unit in ℓ_1 where $D = -\tilde{N}Y_1 + \tilde{M}X_1$ and $Q := -(\tilde{X}Y_1 - \tilde{Y}X_1)D^{-1}$. Therefore D is a unit in ℓ_1. Thus D^{-1} is a stable system and therefore Q is also stable. Multiplyng both sides of the above equation by $\begin{pmatrix} \hat{M} & \hat{Y} \\ \hat{N} & \hat{X} \end{pmatrix}$ we have

$$\begin{pmatrix} \hat{M} & \hat{Y}_1 \\ \hat{N} & \hat{X}_1 \end{pmatrix} = \begin{pmatrix} \hat{M} & (\hat{Y} - \hat{M}\hat{Q})\hat{D} \\ \hat{N} & (\hat{X} - \hat{N}\hat{Q})\hat{D} \end{pmatrix}.$$

By comparing entries in the above equality we have the result that $K = (Y - MQ)(X - NQ)^{-1}$. This proves the theorem. □

Let us assume that in Figure 4.1 G and K are FDLTIC systems. Also assume that a stabilizable and detectable state-space description of is described by

$$G = \begin{pmatrix} G_{11} & G_{22} \\ G_{21} & G_{22} \end{pmatrix} = \left[\begin{array}{c|c} A & B_1 \; B_2 \\ \hline C_1 & D_{11} \; D_{22} \\ C_2 & D_{21} \; D_{22} \end{array}\right].$$

This notation is a convenient way of writing

$$G_{11} = \left[\begin{array}{c|c} A & B_1 \\ \hline C_1 & D_{11} \end{array}\right], \quad G_{12} = \left[\begin{array}{c|c} A & B_2 \\ \hline C_1 & D_{12} \end{array}\right], \quad G_{21} = \left[\begin{array}{c|c} A & B_1 \\ \hline C_2 & D_{21} \end{array}\right]$$

and $G_{22} = \left[\begin{array}{c|c} A & B_2 \\ \hline C_2 & D_{22} \end{array}\right]$.

Lemma 4.2.2. *There exists a FDLTIC system K which stabilizes the closed loop in Figure 4.1 if and only if (A, B_2) is stabilizable and (A, C_2) is detectable. If F and L are such that $\rho(A + B_2F) < 1$ and $\rho(A + LC_2) < 1$ then a controller with a state space realization given by*

$$K = \left[\begin{array}{c|c} A + B_2F + LC_2 + LD_{22}F & -L \\ \hline F & 0 \end{array}\right], \tag{4.9}$$

stabilizes the closed loop system depicted in Figure 4.1.

Proof. ($\Leftarrow$) If (A, B_2) is stabilizable and (A, C_2) is detectable then there exist matrices F and L are such that $\rho(A + B_2F) < 1$ and $\rho(A + LC_2) < 1$. Let K be a controller with a state space realization given in (4.9). It can be shown that the closed loop system has a state-space description given by $\left[\begin{array}{c|c} \tilde{A} & \tilde{B} \\ \hline \tilde{C} & \tilde{D} \end{array}\right]$ where

$$\tilde{A} = \begin{pmatrix} A & B_2F \\ -LC_2 & A + B_2F + LC_2 \end{pmatrix},$$

which has the same eigenvalues as the matrix

$$\begin{pmatrix} A + LC_2 & 0 \\ -LC_2 & A + B_2F \end{pmatrix}.$$

Thus $\rho(\tilde{A}) < 1$ from which it follows from Theorem 4.1.2 that the closed loop map is stable.

($\Rightarrow$) If (A, B_2) is not stabilizable or (A, C_2) is not detectable then some eigenvalues of $\tilde{A}$ will remain outside the unit disc for any FDLTIC controller K. Details are left to the reader. □

The controller K given above is called the Luenberger controller

Lemma 4.2.3. *Suppose (A, B_2) is stabilizable and (A, C_2) is detectable. Then FDLTIC system K stabilizes the closed loop system depicted in Figure 4.1 if and only if it stabilizes the closed loop system depicted in Figure 4.2.*

Proof. ($\Rightarrow$) The closed loop map depicted in Figure 4.1 is described by the equations

$$\begin{aligned} z &= G_{11}w + G_{12}u \\ y &= G_{21}w + G_{22}u \\ u &= Ky + Kv_2 + v_1. \end{aligned} \tag{4.10}$$

The description of the closed loop map depicted in Figure 4.2 is given by

$$\begin{aligned} y &= G_{22}u \\ u &= Ky + Kv_2 + v_1. \end{aligned} \tag{4.11}$$

It is thus clear (substitute $w = 0$ in (4.10)) that if the map from $(w,\ v_1,\ v_2)$ to $(z,\ u,\ y)$ in (4.10) is stable then map from $(v_1,\ v_2)$ to $(u,\ y)$ in (4.11) is stable.

($\Leftarrow$) Suppose K is a stabilizing controller for the closed loop map in Figure 4.2. Let $\left[\begin{array}{c|c} A_K & B_K \\ \hline C_K & D_K \end{array}\right]$ be a stabilizable and detectable state-space description of K. By assumption $\left[\begin{array}{c|c} A & B_2 \\ \hline C_2 & D_{22} \end{array}\right]$ is a stabilizable and detectable state-space description of G_{22}. Suppose, $\left[\begin{array}{c|c} \overline{A} & \overline{B} \\ \hline \overline{C} & \overline{D} \end{array}\right]$ is a state-space description of the closed loop map obtained by employing the aforementioned state-space descriptions of G_{22} and K. Then one can show that $(\overline{A}, \overline{B})$ and $(\overline{A}, \overline{C})$ are stabilizable and detectable [2]. Thus from Theorem 4.1.2 it follows that $\rho(\overline{A}) < 1$.

If $\left[\begin{array}{c|c} \tilde{A} & \tilde{B} \\ \hline \tilde{C} & \tilde{D} \end{array}\right]$ is a description of the closed loop map in Figure 4.1 obtained by using the descriptions $\left[\begin{array}{c|c} A_K & B_K \\ \hline C_K & D_K \end{array}\right]$ for K and $\left[\begin{array}{c|c} A & B \\ \hline C & D \end{array}\right]$ for G_{22} then by computing $\tilde{A}$ one can verify that $\tilde{A} = \overline{A}$. Thus $\rho(\tilde{A}) < 1$ and therefore from Theorem 4.1.2 it follows that the closed loop system in Figure 4.1 is stable. □

Theorem 4.2.2. *Suppose (A, B_2) is stabilizable and (A, C_2) is detectable. Let FDLTIC system G_{22} admit a dcf as given in Lemma 4.2.1. Then K is a FDLTIC stabilizing controller for the closed loop system in Figure 4.1 if and only if*

$$\hat{K} = (\hat{Y} - \hat{M}\hat{Q})(\hat{X} - \hat{N}\hat{Q})^{-1} = (\hat{\tilde{X}} - \hat{Q}\hat{\tilde{N}})^{-1}(\hat{\tilde{Y}} - \hat{Q}\hat{\tilde{M}}),$$

for some FDLTIC stable system Q.

Proof. Follows immediately from Theorem 4.2.1 and Lemma 4.2.3. □

By using the above parametrization we can show that

$$K(I - G_{22}K)^{-1} = (Y - MQ)\tilde{M}.$$

The map from w to z in Figure 4.1 is given by

$$\Phi = G_{11} + G_{12}K(I - G_{22}K)^{-1}G_{21}.$$

Thus we have the folowing theorem

Theorem 4.2.3. *Let G be FDLTIC system and let G_{22} admit a dcf as given in Lemma 4.2.1. Φ is a map from w to z in Figure 4.1 for some FDLTIC, K which stabilizes the closed loop if and only if*

$$\Phi = H - UQV,$$

where

$$\begin{aligned} H &= G_{11} + G_{12}Y\tilde{M}G_{21} \\ U &= G_{12}M \\ V &= \tilde{M}G_{21} \end{aligned}$$

and Q is some stable FDLTIC system.

We now present a result which is a generalization of Theorem 4.2.3.

Theorem 4.2.4 (Youla parametrization). *Let G be a FDLTIC system and let G_{22} admit a dcf as given in Lemma 4.2.1. Φ is a map from w to z in Figure 4.1 for some linear, time invariant, causal K which stabilizes the closed loop in the ℓ_∞ sense if and only if*

$$\Phi = H - UQV,$$

where

$$\begin{aligned} H &= G_{11} + G_{12} Y \tilde{M} G_{21} \\ U &= G_{12} M \\ V &= \tilde{M} G_{21} \end{aligned}$$

and Q is some ℓ_∞ stable system.

The parameter Q is often referred to as the Youla parameter. The difference between Theorem 4.2.3 and Theorem 4.2.4 is that in Theorem 4.2.4 the controller K is not restricted to be finite-dimensional. The proof of this theorem is similar to the one presented for Theorem 4.2.3 except that an analogous result for coprime factorization over ℓ_∞ stable systems is utilized.

5. SISO $\ell_1/\ \mathcal{H}_2$ Problem

Consider the system of Figure 5.1 where $w = (w_1\ w_2)$ is the exogenous disturbance, $\overline{z} = (z_1\ z_2)$ is the regulated output, $\overline{u}$ is the control input and y is the measured output. A number of control design problems can be cast in the framework depicted by Figure 5.1 (see [3]). In feedback control design the objective is to design a controller, K such that with $\overline{u} = Ky$ the resulting closed loop map $\Phi_{\overline{z}w}$ from w to $\overline{z}$ is stable (see Figure 5.1) and satisfies certain performance criteria. Such criteria may be posed in terms of a measure on $\Phi_{\overline{z}w}$ which depends on the signal norms of w and $\overline{z}$ that may be of interest in a particular situation.

For example, the $\mathcal{H}_\infty$ norm of the closed loop measures the energy of the regulated output $\overline{z}$ for the worst disturbance w whose energy is bounded by one i.e.

$$||\Phi_{\overline{z}w}||_{\mathcal{H}_\infty} = \sup_{||w||_2 \leq 1} ||\Phi_{\overline{z}w} w||_2.$$

The standard $\mathcal{H}_\infty$ problem minimizes this norm over all achievable closed loop maps. Thus the standard $\mathcal{H}_\infty$ problem solves the following problem:

$$\min_{K \text{ stabilizing}} ||\Phi_{\overline{z}w}(K)||_{\mathcal{H}_\infty},$$

where $\Phi_{\overline{z}w}(K)$ is the closed loop map from w to $\overline{z}$. The two norm of the closed loop, $||\Phi_{\overline{z}w}||_2$ measures the energy in the regulated output $\overline{z}$ for a unit pulse input, w. The standard $\mathcal{H}_2$ problem finds a stabilizing controller K which results in a closed loop map which has the minimum $\mathcal{H}_2$ norm when compared to all other closed loop maps achievable through stabilizing

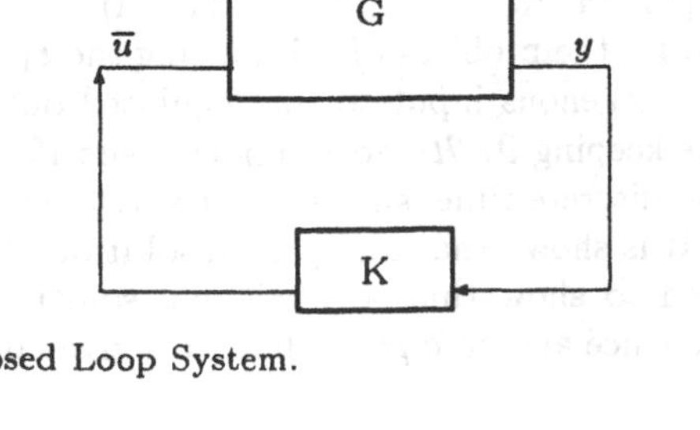

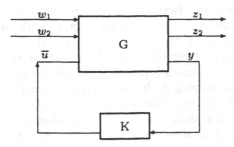

Fig. 5.1. Closed Loop System.

controllers. State space solutions for both the above mentioned problems are provided in [4].

The ℓ_1 norm of the closed loop, $||\Phi_{\overline{z}w}||_1$ is the infinity norm of the regulated output $\overline{z}$, for the worst disturbance w whose magnitude is bounded by one i.e.

$$||\Phi_{\overline{z}w}||_1 = \sup_{||w||_\infty \leq 1} ||\Phi_{\overline{z}w} w||_\infty .$$

The standard ℓ_1 problem finds a controller which minimizes this norm over all closed loop maps that are achevieable through stabilizing controllers. Thus the standard ℓ_1 problem is to determine a controller K which solves the following problem:

$$\min_{K \text{ stabilizing}} ||\Phi_{\overline{z}w}(K)||_1 .$$

It is shown in [5] that this problem reduces to a finite dimensional linear program for the 1-block case.

All of the previous criteria refer to a single performance measure of the closed loop. It is well known (see for example [3]) that minimization with respect to one norm may not necessarily yield good performance with respect to another. This has led researchers to consider problems where multiple measures of the closed loop are incorporated directly into the design. One of the important classes of problems in this category is the mixed $\mathcal{H}_2/\ \mathcal{H}_\infty$ design. Here, the focus is on problems which include the $\mathcal{H}_2$ and the $\mathcal{H}_\infty$ norms of the closed loop in their definitions. Several state space results are available in this class (e.g., [6, 7]). Another important class of problems considers the interplay between the ℓ_1 and the $\mathcal{H}_\infty$ norm of the closed loop. Problems from this class are addressed in [8, 9, 10, 11, 12].

In this book we will design controllers which guarantee performance as reflected by the $\mathcal{H}_2$ and the ℓ_1 measures. In this and the next chapter we present single-input single-output (SISO) (where there is only one exogenous disturbance and only one regulated variable) problems for this class. In Chapter 7 and Chapter 8 we present the multi-input multi-output (MIMO) problems where we study the interplay between the $\mathcal{H}_2$ and the ℓ_1 norms for a general plant. The SISO problems serve to highlight the characteristics of the $\mathcal{H}_2 - \ell_1$ problems. The relevant literature for various methodologies which address the objective of the $\mathcal{H}_2$ measure in conjunction with time domain measures can be found in [13, 14, 15, 16, 17, 18, 19, 20, 21].

In this chapter we consider the problem of minimizing the ℓ_1 norm of the transfer function from the exogenous input to the regulated output over all stabilizing controllers while keeping its $\mathcal{H}_2$ norm under a specified level. The problem is analysed for the discrete-time, single-input single-output (SISO), linear time invariant case. It is shown that an optimal solution always exists. Duality theory is employed to show that any optimal solution is a finite impulse response (FIR) sequence and an *a priori* bound is given on its length.

Thus, the problem can be reduced to a finite dimensional convex optimization problem with an *a priori* determined dimension. Finally it is shown that, in the region of interest of the $\mathcal{H}_2$ constraint level the optimal is unique and continuous with respect to changes in the constraint level.

5.1 Problem Formulation

Consider the standard feedback problem represented in Figure 5.2 where P and K are the plant and the controller respectively. Let w represent the exogenous input, $\overline{z}$ represent the output of interest, y is the measured output and $\overline{u}$ is the control input where $\overline{z}$ and w are assumed scalar. Let ϕ be the closed loop map which maps $w \rightarrow \overline{z}$. From the Youla parametrization (see Theorem 4.2.4) it is known that all achievable closed loop maps under stabilizing controllers are given by $\hat{\phi} = \hat{h} - \hat{u}\hat{q}$, where $h, u, q \in \ell_1$; h, u depend only on the plant P and q is a free parameter in ℓ_1. The following is a result from complex analysis

Theorem 5.1.1. *Given a sequence $\{t(i)\}$ in ℓ_1 with the associated λ transform $\hat{t}(\lambda)$ then the following statements are equivalent.*

1. $\frac{d^k \hat{t}}{d\lambda^k}|_{\lambda=\lambda_0} = 0$ *for* $k = 0, 1, \ldots, \sigma - 1$.
2. $\hat{t}(\lambda) = (\lambda - \lambda_0)^{\sigma-1}\hat{s}(\lambda)$ *where $\hat{s}$ is the λ-transform of some element* $s \in \ell_1$.

Throughout the chapter we make the following assumption.

Assumption 1 *All the zeros of $\hat{u}$ (the λ transform of u) inside the unit disc are real and distinct. Also, $\hat{u}$ has no zeros on the unit circle.*

The assumption that all zeros of $\hat{u}$ which are inside the open unit disc are real and distinct is not restrictive and is made to streamline the presentation of the chapter. Let the zeros of u which are inside the unit disc be given by $z_1, z_2, \ldots, z_n$. Let

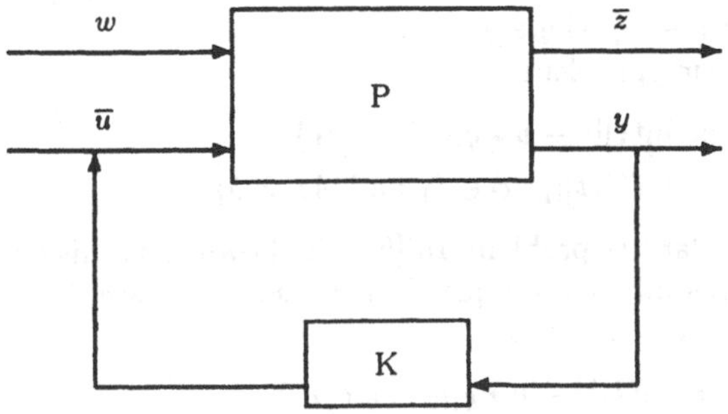

Fig. 5.2. Plant controller configuration

$$\Theta := \{\phi : \text{ there exists } q \in \ell_1 \text{ with } \hat{\phi} = \hat{h} - \hat{u}\hat{q}\}.$$

Θ is the set of all achievable closed loop maps under stabilizing controllers. Let $A : \ell_1 \to R^n$ be given by

$$A = \begin{pmatrix} 1 & z_1 & z_1^2 & z_1^3 & \dots \\ 1 & z_2 & z_2^2 & z_2^3 & \dots \\ \vdots & \vdots & \vdots & \vdots & \vdots \\ 1 & z_n & z_n^2 & z_n^3 & \dots \end{pmatrix},$$

and $b \in R^n$ be given by

$$b = \begin{pmatrix} \hat{h}(z_1) \\ \hat{h}(z_2) \\ \vdots \\ \hat{h}(z_n) \end{pmatrix}.$$

Theorem 5.1.2. *The following is true:*

$$\begin{aligned} \Theta &= \{\phi \in \ell_1 : \hat{\phi}(z_i) = \hat{h}(z_i) \text{ for all } i = 1, \dots, n\} \\ &= \{\phi \in \ell_1 : A\phi = b\}. \end{aligned}$$

Proof. ($\Rightarrow$) Suppose $\phi \in \Theta$. Then $\hat{\phi} = \hat{h} - \hat{u}\hat{q}$ for some $q \in \ell_1$. It follows that $\phi \in \ell_1$ and that $\hat{\phi}(z_i) = \hat{h}(z_i) - \hat{u}(z_i)\hat{q}(z_i) = 0$, for all $i = 1, \dots, n$. Thus $\phi \in \ell_1$ and $A\phi = b$.

($\Leftarrow$) Suppose, $\hat{\phi}(z_i) = \hat{h}(z_i)$ for all $i = 1, \dots, n$ and $\phi \in \ell_1$. It follows that $(\hat{h} - \hat{\phi})(z_i) = 0$ for all $i = 1, \dots, n$ and $h - \phi \in \ell_1$. From Theorem 5.1.1 it follows that $(\hat{h} - \hat{\phi})(\lambda) = (\lambda - \lambda_1)\dots(\lambda - \lambda_n)\hat{p}(\lambda)$ where $\hat{p}(\lambda)$ is the λ-transform of an element $p \in \ell_1$. Let $\hat{u}(\lambda) = (\lambda - \lambda_1)\hat{p}_u$ where $\hat{p}_u$ has no zeros inside the closed unit disc in the complex plane. Thus it follows that

$$\hat{q} := \frac{\hat{h} - \hat{\phi}}{\hat{u}} = \frac{\hat{p}}{\hat{p}_u}.$$

As $\hat{p}_u$ has no zeros inside the unit disc it is a unit in ℓ_1. It follows that $q \in \ell_1$. Thus $\hat{\phi} = \hat{h} - \hat{u}\hat{q}$ where $q \in \ell_1$. □

The following problem

$$\begin{aligned} \nu_\infty &:= \inf\{||h - u * q||_1 : q \in \ell_1\} \\ &= \inf\{||\phi||_1 : \phi \in \ell_1 \text{ and } A\phi = b\}, \end{aligned} \tag{5.1}$$

is the standard ℓ_1 problem. In [5] it is shown that this problem has a solution which is possibly non-unique. Optimal solutions are shown to be finite impulse response sequences. Let

$$\begin{aligned} \mu_\infty &:= \inf\{||h - u * q||_2{}^2 : q \in \ell_1\}, \\ &= \inf\{||\phi||_2{}^2 : \phi \in \ell_1 \text{ and } A\phi = b\}, \end{aligned} \tag{5.2}$$

which is the standard $\mathcal{H}_2$ problem. The solution to this problem is unique and the solution is an infinite impulse response sequence. Define

$$m_1 := \inf_{A\phi=b, ||\phi||_2^2 \leq \mu_\infty} ||\phi||_1, \tag{5.3}$$

which is the ℓ_1 norm of the unique optimal solution of the standard $\mathcal{H}_2$ problem. Let

$$m_2 := \inf_{A\phi=b, ||\phi||_1 \leq \nu_\infty} ||\phi||_2^2, \tag{5.4}$$

which is the infimum over the ℓ_2 norms of the optimal solutions of the standard ℓ_1 problem.

The problem of interest is : *Given a positive constant $\gamma > \mu_\infty$ obtain a solution to the following mixed objective problem:*

$$\begin{aligned}\nu_\gamma &:= \inf\{||h - u * q||_1 : q \in \ell_1 \text{ and } < h - u * q, h - u * q > \leq \gamma\} \\ &= \inf\{||\phi||_1 : \phi \in \ell_1 \text{ , } A\phi = b \text{ and } < \phi, \phi > \leq \gamma\}.\end{aligned} \tag{5.5}$$

In the following sections we will study this problem from the point of view of existence, structure, continuity, and computation of the optimal solutions.

5.2 Optimal Solutions and their Properties

In the first part of this section we show that (5.5) always has a solution. In the second part we show that any solution to (5.5) is of finite length and in the third we give an *a priori* bound on the length.

5.2.1 Existence of a Solution

Here we show that a solution to (5.5) always exists.

Theorem 5.2.1. *There exists $\phi_0 \in \Phi$ such that*

$$||\phi_0||_1 = \inf_{\phi \in \Phi} \{||\phi||_1\},$$

where $\Phi := \{\phi \in \ell_1 : A\phi = b$ and $< \phi, \phi > \leq \gamma\}$ with $\gamma > \mu_\infty$. Therefore the infimum in (5.5) is a minimum.

Proof. We denote the feasible set of our problem by $\Phi := \{\phi \in \ell_1 : A\phi = b$ and $< \phi, \phi > \leq \gamma\}$. $\nu_\gamma < \infty$ because $\gamma > \mu_\infty$ and therefore the feasible set is not empty. Let $B := \{\phi \in \ell_1 : ||\phi||_1 \leq \nu_\gamma + 1\}$. It is clear that

$$\nu_\gamma = \inf_{\phi \in \Phi \cap B} \{||\phi||_1\}.$$

Therefore given $i > 0$ there exists $\phi_i \in \Phi \cap B$ such that $||\phi_i||_1 \leq \nu_\gamma + \frac{1}{i}$. B is a bounded set in $\ell_1 = c_0^*$. It follows from the Banach-Alaoglu result (see Theorem 2.3.1) that B is $W(c_0^*, c_0)$ compact. Using the fact that c_0 is separable we know (see Theoerm 2.3.2) that there exists a subsequence $\{\phi_{i_k}\}$ of $\{\phi_i\}$ and $\phi_0 \in \Phi \cap B$ such that $\{\phi_{i_k}\} \to \phi_0$ in the $W(c_0^*, c_0)$ sense, that is for all v in c_0

$$< v, \phi_{i_k} > \to < v, \phi_0 > \text{ as } k \to \infty. \tag{5.6}$$

Let the jth row of A be denoted by a_j and the j^{th}element of b be given by b_j . Then as $a_j \in c_0$ we have,

$$< a_j, \phi_{i_k} > \to < a_j, \phi_0 > \text{ as } k \to \infty \text{ for all } j = 1, 2, \ldots, n. \tag{5.7}$$

As $A(\phi_{i_k}) = b$ we have $< a_j, \phi_{i_k} >= b_j$ for all k and for all j which implies $< a_j, \phi_0 >= b_j$ for all j. Therefore we have $A(\phi_0) = b$. As $l_2 \subset c_0$ we have from (5.6) that for all v in l_2

$$< v, \phi_{i_k} > \to < v, \phi_0 > \text{ as } k \to \infty, \tag{5.8}$$

which shows that $\phi_{i_k} \to \phi_0$ in $W(l_2^*, l_2)$. Also, from the construction of ϕ_{i_k} we know that $||\phi_{i_k}||_2 \leq \sqrt{\gamma}$. From Theorem 2.3.1 (Banach-Alaoglu theorem) we conclude that $< \phi_0, \phi_0 > \leq \gamma$ and therefore we have shown that $\phi_0 \in \Phi$. Recall that ϕ_{i_k} were chosen so that $||\phi_{i_k}||_1 \leq \nu_\gamma + \frac{1}{i_k}$. From Theorem 2.3.1 we have that $||\phi_0||_1 \leq \nu_\gamma + \frac{1}{i_k}$ for all k. Therefore $||\phi_0||_1 \leq \nu_\gamma$. As $\phi_0 \in \Phi$ (which is the feasible set) we have $||\phi_0||_1 = \nu_\gamma$. This proves the theorem. □

5.2.2 Structure of Optimal Solutions

In this subsection we use duality results to show that every optimal solution is of finite length. The following two lemmas establish the dual problem.

Lemma 5.2.1.

$$\nu_\gamma = \max\{\varphi(y_1, y_2) : y_1 \geq 0 \textit{ and } y_2 \in R^n\}, \tag{5.9}$$

where

$$\varphi(y_1, y_2) := \inf_{\phi \in \ell_1} \{||\phi||_1 + y_1(< \phi, \phi > -\gamma) + < b - A\phi, y_2 >\}.$$

Proof. We will apply Theorem 3.3.2 (Kuhn-Tucker-Lagrange duality theorem) to get the result. Let X, Ω, Y, Z in Theorem 3.3.2 correspond to ℓ_1, ℓ_1, R^n and R respectively. Let $g(\phi) :=< \phi, \phi > -\gamma$, $H(\phi) := b - A\phi$. With this notation we have $Z^* = R$.

A has full range which implies $0 \in int[range(H)]$. $\gamma > \mu_\infty$ and therefore their exists ϕ_1 such that $< \phi_1, \phi_1 > -\gamma < 0$ and $H(\phi_1) = 0$. Therefore all the conditions of Theorem 3.3.2 are satisfied. From Theorem 3.3.2 (Kuhn-Tucker-Lagrange duality theorem) we have

$$\nu_\gamma = \max_{y_1 \geq 0, y_2 \in R^n} \inf_{\phi \in \ell_1} \{||\phi||_1 + y_1(< \phi, \phi > -\gamma) + < b - A\phi, y_2 >\}.$$

This proves the lemma. □

The right hand side of (5.9) is the *dual problem*.

Lemma 5.2.2. *The dual problem is given by :*

$$\max\{\varphi(y_1, y_2) : y_1 \geq 0 \quad and \quad y_2 \in R^n\}, \tag{5.10}$$

where $\varphi(y_1, y_2)$ *is*

$$\inf_{\phi \in \ell_1, \phi(i)v(i) \geq 0} \{||\phi||_1 + y_1(<\phi, \phi> - \gamma) + <b, y_2> - <\phi, v>\}.$$

$v(i)$ *is defined by* $v(i) := A^* y_2(i)$.

Proof. Let $y_1 \geq 0$, $y_2 \in R^n$. It is clear that

$$\begin{aligned} &\inf_{\phi \in \ell_1} \{||\phi||_1 + y_1(<\phi, \phi> - \gamma) + <b - A\phi, y_2>\} \\ &= \inf_{\phi \in \ell_1} \{||\phi||_1 + y_1(<\phi, \phi> - \gamma) + <b, y_2> - <\phi, v>\}. \end{aligned}$$

Suppose $\phi \in \ell_1$ and there exists i such that $\phi(i)v(i) < 0$ then define $\phi^1 \in \ell_1$ such that $\phi^1(j) = \phi(j)$ for all $j \neq i$ and $\phi^1(i) = 0$. Therefore we have $||\phi||_1 + y_1(<\phi, \phi> - \gamma) + <b, y_2> - <\phi, v> \geq ||\phi^1||_1 + y_1(<\phi^1, \phi^1> - \gamma) + <b, y_2> - <\phi^1, v>$. This shows that we can restrict ϕ in the infimization to satisfy $\phi(i)v(i) \geq 0$. This proves the lemma. □

The following theorem is the main result of this subsection. It shows that any solution of (5.5) is a finite impulse response sequence.

Theorem 5.2.2. *Define* $\mathcal{T}$:=$\{\phi \in \ell_1$: *there exists* L^* *with* $\phi(i) = 0$ *if* $i \geq L^*\}$. *The dual of the problem is given by:*

$$\max\{\varphi(y_1, y_2) : y_1 \geq 0, y_2 \in R^n\}, \tag{5.11}$$

where $\varphi(y_1, y_2)$ *is*

$$\inf_{\phi \in \mathcal{T}, \phi(i)v(i) \geq 0} \{||\phi||_1 + y_1(<\phi, \phi> - \gamma) + <b, y_2> - <\phi, v>\}.$$

$v(i)$ *defined by* $v(i) = A^* y_2(i)$. *Also, any optimal solution* ϕ_0 *of (5.5) belongs to* $\mathcal{T}$.

Proof. Let $y_1^\gamma \geq 0$, $y_2^\gamma \in R^n$ be the solution to

$$\max_{y_1 \geq 0, y_2 \in R^n} \inf_{\phi \in \ell_1, \phi(i)v(i) \geq 0} \{||\phi||_1 + y_1(<\phi, \phi> - \gamma) + <b - A\phi, y_2>\}.$$

It is easy to show that there exists L^* such that $v^\gamma(i) := (A^* y_2^\gamma)(i)$ satisfies $|v^\gamma(i)| < 1$ if $i \geq L^*$. If $\phi(i)v^\gamma(i) \geq 0$ for all i then,

$$\begin{aligned} &||\phi||_1 + y_1^\gamma(<\phi, \phi> - \gamma) + <b, y_2^\gamma> - <\phi, v^\gamma> \\ &= \sum_{i=0}^{\infty} \{|\phi(i)| + y_1^\gamma(\phi(i))^2 - \phi(i)v^\gamma(i)\} - y_1^\gamma \gamma + <y_2^\gamma, b> \\ &= \sum_{i=0}^{\infty} \{\phi(i)(sgn(v^\gamma(i)) - v^\gamma(i)) + y_1^\gamma(\phi(i))^2\} - y_1^\gamma \gamma + <y_2^\gamma, b> \\ &= \sum_{i=0}^{L^*} \{\phi(i)(sgn(v^\gamma(i)) - v^\gamma(i)) + y_1^\gamma(\phi(i))^2\} \end{aligned}$$

$$+\sum_{i=L^*+1}^{\infty}\{\phi(i)(sgn(v^\gamma(i))-v^\gamma(i))+y_1^\gamma(\phi(i))^2\}-y_1^\gamma\gamma+<y_2^\gamma,b>.$$

Suppose $|v^\gamma(i)|<1$. Then we have,

$$\phi(i)(sgn(v^\gamma(i))-v^\gamma(i))+y_1^\gamma(\phi(i))^2\geq 0$$

and equals zero only if $\phi(i)=0$. Therefore, in the infimization we can restrict $\phi(i)=0$ whenever $|v^\gamma(i)|<1$. As $|v^\gamma(i)|<1$ for all $i\geq L^*$ it follows that we can restrict ϕ to $\mathcal{T}$ in the infimization. In Theorem 5.2.1 we showed that there exists a solution ϕ_0 to the primal. From Theorem 3.3.2 (Kuhn-Tucker-Lagrange duality theorem) we have that ϕ_0 minimizes

$$||\phi||_1+y_1^\gamma(<\phi,\phi>-\gamma)+<b,y_2^\gamma>-<\phi,v^\gamma>,$$

over all ϕ in ℓ_1. From the previous discussion it follows that $\phi_0\in\mathcal{T}$. This proves the theorem. □

5.2.3 An A priori Bound on the Length of Any Optimal Solution

In this subsection we give an *a priori* bound on the length of any solution to (5.5). First we establish the following three lemmas.

Lemma 5.2.3. *Let* $\gamma>\mu_\infty$, $m_1:=\inf_{A\phi=b,<\phi,\phi>\leq\mu_\infty}||\phi||_1$, *and* $\nu_\gamma:=\inf_{A\phi=b,<\phi,\phi>\leq\gamma}||\phi||_1$. *Let* y_1^γ,y_2^γ *represent a dual solution as obtained in (5.10). Then* $y_1^\gamma\leq M_\gamma$ *where* $M_\gamma:=\frac{m_1}{\gamma-\mu_\infty}$.

Proof. Let $\gamma>\gamma_1>\mu_\infty$ and $\nu_{\gamma_1}:=\inf_{A\phi=b,<\phi,\phi>\leq\gamma_1}||\phi||_1$. Let y_1^γ,y_2^γ represent a dual solution as obtained in (5.10). From Corollary 3.3.1 (a sensitivity result) we have

$$<\gamma-\gamma_1,y_1^\gamma>\leq\nu_{\gamma_1}-\nu_\gamma\leq\nu_{\gamma_1}\leq m_1,$$

which implies that $y_1^\gamma\leq\frac{m_1}{\gamma-\gamma_1}$. This holds for all $\gamma>\gamma_1>\mu_\infty$. Therefore $M_\gamma:=\frac{m_1}{\gamma-\mu_\infty}$ is an *a priori* bound on y_1^γ. This proves the lemma. □

Lemma 5.2.4. *Let* ϕ_0 *be a solution of the primal (5.5). Let* y_1^γ,y_2^γ *represent the corresponding dual solution as obtained in (5.10). Let* $v^\gamma:=A^*y_2^\gamma$ *then,*

$$\begin{aligned}y_1^\gamma\phi_0(i)&=\tfrac{v^\gamma(i)-1}{2}\quad if\ v^\gamma(i)>1\\&=\tfrac{v^\gamma(i)+1}{2}\quad if\ v^\gamma(i)<-1\\&=\quad 0\qquad if\ |v^\gamma(i)|\leq 1.\end{aligned}$$

Also, $||v^\gamma||_\infty\leq\alpha_\gamma$ *where* $\alpha_\gamma=\frac{2m_1\sqrt{\gamma}}{\gamma-\mu_\infty}+1$.

Proof. Let

$$L(\phi) := \sum_{i=0}^{\infty}\{\phi(i)(sgn(v^{\gamma}(i)) - v^{\gamma}(i)) + y_1^{\gamma}(\phi(i))^2\} - \gamma y_1^{\gamma} + < b, y_2^{\gamma} > .$$

Suppose $|v^{\gamma}(i)| = 1$. Now, if $y_1^{\gamma} = 0$ then it is clear that $y_1^{\gamma}\phi_0(i) = 0$. If $y_1^{\gamma} > 0$ then as ϕ_0 minimizes $L(\phi)$ we have $\phi_0(i) = 0$. We have already shown that if $|v^{\gamma}(i)| < 1$ then $\phi_0(i) = 0$. Therefore, $y_1^{\gamma}\phi_0(i) = 0$ if $|v^{\gamma}(i)| \leq 1$.

Suppose $v^{\gamma}(i) > 1$ then it is easy to show that there exists $\phi(i)$ such that $\phi(i) \geq 0$ and $\phi(i)(sgn(v^{\gamma}(i)) - v^{\gamma}(i)) + y_1^{\gamma}(\phi(i))^2 < 0$. As any optimal minimizes $L(\phi)$ we know that $\phi_0(i)(sgn(v^{\gamma}(i)) - v^{\gamma}(i)) + y_1^{\gamma}(\phi_0(i))^2 < 0$, which implies $\phi_0(i) > 0$ and therefore $1 - v^{\gamma}(i) + 2y_1^{\gamma}\phi_0(i) = 0$. This implies that $y_1^{\gamma}\phi_0(i) = \frac{v^{\gamma}(i)-1}{2}$. Similarly the result follows when $v^{\gamma}(i) < -1$. Therefore, $||v^{\gamma}||_{\infty} \leq 2M_{\gamma}||\phi_0||_{\infty} + 1 \leq \frac{2m_1}{\gamma - \mu_{\infty}}\, ||\phi_0||_{\infty} + 1 \leq \frac{2m_1\sqrt{\gamma}}{\gamma-\mu_{\infty}} + 1$. The last inequality follows from the fact that $< \phi_0, \phi_0 > \leq \gamma$. This implies that $\alpha_{\gamma} := \frac{2m_1\sqrt{\gamma}}{\gamma-\mu_{\infty}} + 1$ is an *a priori* upper bound on $||v^{\gamma}||_{\infty}$. This proves the lemma. □

Lemma 5.2.5. *If $y_2 \in R^n$ is such that $||A^*y_2||_{\infty} \leq \alpha_{\gamma}$ then there exists a positive integer L^* independent of y_2 such that $|(A^*y_2)(i)| < 1$ for all $i \geq L^*$.*

Proof. Define

$$A_L^* = \begin{pmatrix} 1 & 1 & 1 & \dots & 1 \\ z_1 & z_2 & z_3 & \dots & z_n \\ \vdots & \vdots & \vdots & \vdots & \vdots \\ z_1^L & z_2^L & z_3^L & \dots & z_n^L \end{pmatrix},$$

$A_L^* : R^n \to R^{L+1}$. With this definition we have $A_{\infty}^* = A^*$. Let $y_2 \in R^n$ be such that $||A^*y_2||_{\infty} \leq \alpha_{\gamma}$. Choose any L such that $L \geq (n-1)$. As $z_i, i = 1, \dots, n$ are distinct A_L^* has full column rank. A_L^* can be regarded as a linear map taking $(R^n, ||.||_1) \to (R^{L+1}, ||.||_{\infty})$. As A_L^* has full column rank we can define the left inverse of A_L^*, $(A_L^*)^{-l}$ which takes $(R^{L+1}, ||.||_{\infty}) \to (R^n, ||.||_1)$. Let the induced norm of $(A_L^*)^{-l}$ be given by $||\,(A_L^*)^{-l}\,||_{\infty,1}$. $y_2 \in R^n$ is such that $||A^*y_2||_{\infty} \leq \alpha_{\gamma}$ and therefore $||A_L^*y_2||_{\infty} \leq \alpha_{\gamma}$. It follows that,

$$||y_2||_1 \leq ||\,(A_L^*)^{-l}\,||_{\infty,1}\, ||A_L^*y_2||_{\infty} \leq ||\,(A_L^*)^{-l}\,||_{\infty,1}\, \alpha_{\gamma}. \tag{5.12}$$

Choose L^* such that

$$\max_{k=1,\dots,n} |z_k|^{L^*}\ ||\,(A_L^*)^{-l}\,||_{\infty,1}\, \alpha_{\gamma} < 1. \tag{5.13}$$

There always exists such an L^* because $|z_k| < 1$ for all $k = 1, \dots, n$. Note that L^* does not depend on y_2. For any $i \geq L^*$ we have

$$
\begin{aligned}
|(A^* y_2)(i)| = |\sum_{k=1}^{k=n} z_k^i y_2(k)| &\leq \max_{k=1,\ldots,n} |z_k|^i ||y_2||_1 \\
&\leq \max_{k=1,\ldots,n} |z_k|^i \ || (A_L^*)^{-l} ||_{\infty,1} \alpha_\gamma \\
&\leq \max_{k=1,\ldots,n} |z_k|^{L^*} \ || (A_L^*)^{-l} ||_{\infty,1} \alpha_\gamma .
\end{aligned}
$$

The second inequality follows from (5.12). From (5.13) we have $|(A^* y_2)(i)| < 1$ if $i \geq L^*$. This proves the lemma. □

The following theorem is the main result of the section.

Theorem 5.2.3. *Every solution ϕ_0 of the primal (5.5) is such that $\phi(i) = 0$ if $i \geq L^*$ where L^* given in Lemma 5.2.5 can be determined a priori. Furthermore the upperbound on lengths of the optimal solutions is nonincreasing as a function of γ.*

Proof. Using Lemma 5.2.4 we can bound on $||v^\gamma||_\infty$ by α_γ. By applying Lemma 5.2.5 we conclude that there exists L_γ^* (which can be determined a priori) such that $|v^\gamma(i)| < 1$ if $i \geq L_\gamma^*$. Using the fact that $\phi_0(i) = 0$ if $|v^\gamma(i)| < 1$ we conclude that $\phi_0 = 0$ if $i \geq L_\gamma^*$. L_γ^* was chosen to satisfy

$$
\max_{k=1,\ldots,n} |z_k|^{L^*} \ || (A_L^*)^{-l} ||_{\infty,1} \alpha_\gamma < 1.
$$

α_γ is nonincreasing as a function of γ . Therefore L_γ^* is nonincresaing as a function of γ. This proves the theorem. □

Note that as $\alpha_\gamma = \frac{2m_1\sqrt{\gamma}}{\gamma - \mu_\infty} + 1$, we have that the upper bound on lengths of the solutions increases to infinity as γ decreases to μ_∞. This is commmensurate with the fact that the optimal solution for the standard $\mathcal{H}_2$ problem (5.2) is an infinite impulse response sequence.

The above theorem shows that the problem at hand is a finite dimensional convex problem of *a priori* determined dimension. In particular, in view of Theorem 5.2.3 the problem that needs to be solved is as follows

$$
\nu_\gamma = \min_{A_{L^*}\phi = b, <\phi,\phi> \leq \gamma} \sum_{k=0}^{L^*} |\phi(k)|, \tag{5.14}
$$

where $A_{L^*} = \begin{pmatrix} 1 & z_1 & z_1^2 & \ldots & z_1^{L^*} \\ 1 & z_2 & z_2^2 & \ldots & z_2^{L^*} \\ \vdots & \vdots & \vdots & \vdots & \vdots \\ 1 & z_n & z_n^2 & \ldots & z_n^{L^*} \end{pmatrix}$, and L^* is given in Lemma 5.2.5. An alternative representation can be given as the following lemma suggests

Lemma 5.2.6. *The primal is given by:*

$$
minimize \sum_{k=0}^{L^*} \phi^+(k) + \phi^-(k) \tag{5.15}
$$

$$\begin{aligned} \textit{subject to} \quad & A_{L^*}(\phi^+ - \phi^-) = b \\ & < \phi^+ - \phi^-, \phi^+ - \phi^- > \leq \gamma \\ & \phi^+, \phi^- \textit{ in } R^{L^*} \textit{ with } \phi^+, \phi^- \geq 0. \end{aligned}$$

Proof. Note that in the above theorem the ordering is componentwise for the inequalities. We will show that (5.14) is equivalent to (5.15). Let p_0 denote the value attained by the objective functional in (5.15). Suppose ϕ^+, ϕ^- satisfy the constraints of 5.15. Let $\phi := \phi^+ - \phi^-$. Then it is clear that ϕ satisfies the constraints of (5.14). Also, for each k, $|\phi(k)| = |\phi^+ - \phi^-| \leq |\phi^+| + |\phi^-| = \phi^+(k) + \phi^-(k)$. This implies that $\nu_\gamma \leq p_0$.

Suppose ϕ satisfies the constraints of (5.14). Define ϕ^+ such that $\phi^+(k) = \phi(k)$ if $\phi(k) \geq 0$ and 0 otherwise. Similarly, define ϕ^- such that $\phi^-(k) = -\phi(k)$ if $\phi(k) \leq 0$ and 0 otherwise. It is clear that $\phi = \phi^+ - \phi^-$ and that ϕ^+, ϕ^- satisfy the constraints of (5.15). Also, $|\phi(k)| = \phi^+(k) + \phi^-(k)$. This proves that $\nu_\gamma \geq p_0$. Therefore $\nu_\gamma = p_0$. It is easy to show that if ϕ_0^+, ϕ_0^- is optimal for (5.15) then $\phi_0 := \phi_0^+ - \phi_0^-$ is optimal for (5.14). This proves the lemma. □

This class of convex problems can be solved efficiently using standard methods [22].

5.3 Uniqueness and Continuity of the Solution

In this section we address the issue of uniqueness and continuity of solutions to the primal problem with respect to changes in the constraint level on the $\mathcal{H}_2$ norm of the closed loop map. In the first part we address the issue of uniqueness and in the second part we show that the optimal solution is continuous in the region where it is unique.

5.3.1 Uniqueness of the Optimal Solution

The following three lemmas are established before the main result of this subsection

Lemma 5.3.1. *Let $y_1^\gamma \geq 0$, $y_2^\gamma \in R^n$ be a solution to (5.10). If $y_1^\gamma = 0$ then $\nu_\gamma = \nu_\infty$. This implies that (5.5) reduces to solving a standard ℓ_1 problem.*

Proof. Let $v := A^* y_2$ and ϕ^1 be such that $A\phi^1 = b$. If $y_1^\gamma = 0$ then the dual (5.10) is given by:

$$\max_{y_2 \in R^n} \inf_{\phi(i)v(i) \geq 0} \{||\phi||_1 + < b - A\phi, y_2 >\}$$

$$= \max_{y_2 \in R^n} \inf_{\phi(i)v(i) \geq 0} \sum_{i=0}^{\infty} \{\phi(i)(sgn(v(i)) - v(i))\} + < \phi^1, v >$$

$$= \max_{v \in Range(A^*)} \inf_{\phi(i)v(i) \geq 0} \sum_{i=0}^{\infty} \{\phi(i)(sgn(v(i)) - v(i))\} + < \phi^1, v > .$$

Suppose $||v||_\infty > 1$ then there exists j such that $|v(j)| > 1$. Thus we can choose $\phi(j)$ with $\phi(j)v(j) \geq 0$ such that $\phi(j)(sgn(v(j)) - v(j)) < M$ for any M. This implies that

$$\inf_{\phi(i)v(i) \geq 0} \sum_{i=0}^{\infty} \{\phi(i)(sgn(v(i)) - v(i))\} + < \phi^1, v >= -\infty.$$

Therefore we can restrict v in the maximization to satisfy $||v||_\infty \leq 1$. From arguments similar to that of the proof of Theorem 5.2.3 $\phi(i) = 0$ whenever $|v(i)| < 1$. Therefore the infimum term is zero whenever $||v||_\infty \leq 1$. This implies that the dual problem reduces to:

$$\max_{v \in Range(A^*), ||v||_\infty \leq 1} < \phi^1, v >,$$

which is the same as the dual of the standard ℓ_1 problem as given in (5.1) [3]. This proves the lemma. □

Lemma 5.3.2. *Let $y_1^\gamma \geq 0$, $y_2^\gamma \in R^n$ be a solution in (5.10). If $y_1^\gamma > 0$ then the solution ϕ_0 of (5.5) is unique.*

Proof. Let $L(\phi) := ||\phi||_1 + y_1^\gamma(< \phi, \phi > -\gamma) + < b - A\phi, y_2^\gamma >$. From Theorem 3.3.2 (Kuhn-Tucker-Lagrange duality theorem) we know that ϕ_0 minimizes $L(\phi)$, $\phi \in \ell_1$. If $y_1^\gamma > 0$ then it is easy to show that $L(\phi)$ is strictly convex in ℓ_1. From the Lemma 3.3.2 it follows that ϕ_0 is unique. This proves the lemma. □

The main result of this subsection is now presented.

Theorem 5.3.1. *Define $S := \{\phi : A\phi = b$ and $||\phi||_1 = \nu_\infty\}$, $m_2 := \inf_{\phi \in S} < \phi, \phi >$. The following is true:*

1) *If $\gamma \geq m_2$ then problem (5.5) is equivalent to the standard ℓ_1 problem whose solution is possibly nonunique.*

2) *If $\mu_\infty < \gamma < m_2$ then the solution to (5.5) is unique.*

Proof. Suppose $m_2 < \gamma$ then there exists $\phi_1 \in \ell_1$ such that $A\phi_1 = b$, $||\phi_1||_1 = \nu_\infty$ and $< \phi_1, \phi_1 > \leq \gamma$. This implies that $\nu_\gamma = \inf_{A\phi = b, <\phi,\phi> \leq \gamma} ||\phi||_1 \leq \nu_\infty$. The other inequality is obvious. This proves 1).

Let $\mu_\infty < \gamma < m_2$ and suppose $y_1^\gamma = 0$ then we have shown in Lemma 5.3.1 that $\nu_\gamma = \nu_\infty$. Therefore there exists $\phi_1 \in \ell_1$ such that $||\phi_1||_1 = \nu_\infty$, $A\phi_1 = b$ and $< \phi_1, \phi_1 > \leq \gamma < m_2$. This implies that $\phi_1 \in S$ and $< \phi_1, \phi_1 > < m_2$ which is a contradiction. Therefore $y_1^\gamma > 0$. From Lemma 5.3.2 we know that ϕ_0 is unique. This proves 2). □

The above theorem shows that in the region where the constraint level on the $\mathcal{H}_2$ is essentially of interest (i.e., active) the optimal solution is unique.

5.3.2 Continuity of the Optimal Solution

Following is a theorem which shows that the ℓ_1 norm of the optimal solution is continuous with respect to changes in the constraint level γ.

Theorem 5.3.2. *Let* $\nu_\gamma := \inf_{A\phi=b,<\phi,\phi>\leq\gamma} ||\phi||_1$. *Then* ν_γ *is a continuous function of* γ *on* (μ_∞, ∞).

Proof. If $\gamma \in (\mu_\infty, \infty)$ then it is obvious that $\gamma \in int\{dom(\nu_\gamma)\}$ where $dom(\nu_\gamma) := \{\gamma : -\infty < \nu_\gamma < \infty\}$ is the domain of ν_γ. From Lemma 3.3.3 we know that ν_γ is a convex function of γ. The theorem follows from the fact that every convex function is continuous in the interior of its domain (see Lemma 3.1.2). □

Now we prove that the optimal solution is continuous with respect to changes in the constraint level in the region where the optimal is unique.

Theorem 5.3.3. *Let* $\mu_\infty < \gamma < m_2$. *Let* ϕ_γ *represent the solution of*

$$\nu_\gamma = \min_{A\phi=b,<\phi,\phi>\leq\gamma} ||\phi||_1.$$

Then $\phi_{\gamma_k} \rightarrow \phi_\gamma$ *in the norm topology if* $\gamma_k \rightarrow \gamma$.

Proof. Let $m_1 := \min_{A\phi=b,<\phi,\phi>\leq\mu_\infty} ||\phi||_1$. Then it is obvious that $||\phi_\gamma||_1 = \nu_\gamma \leq m_1$. Without loss of generality assume that $\gamma_k \geq \gamma/2$. Let L^* represent the upper bound on the length of $\phi_{\frac{\gamma}{2}}$. Then as the upperbound is nonincreasing (see Theorem 5.2.3) we can assume that $\phi_{\gamma_k} \in R^{L^*}$. Let $B := \{x : x \in R^{L^*} : ||x||_1 \leq m_1\}$ then we have $\phi_{\gamma_k} \in B$. Therefore there exists a subsequence ϕ_{k_i} of ϕ_{γ_k} and ϕ_1 such that

$$\phi_{k_i} \rightarrow \phi_1 \text{ as } i \rightarrow \infty \text{ in } (R^{L^*}, ||.||_1). \tag{5.16}$$

It is clear as in the proof of Theorem 5.2.1 that $A\phi_1 = b$ as $A\phi_{k_i} = b$ for all i. Also,

$$||\phi_1||_2^2 \leq ||\phi_1 - \phi_{k_i}||_2^2 + ||\phi_{k_i}||_2^2 \leq ||\phi_1 - \phi_{k_i}||_2^2 + \gamma_{k_i}.$$

Taking limits on both sides as $i \rightarrow \infty$ we get $< \phi_1, \phi_1 > \leq \gamma$. This implies that ϕ_1 is a feasible element in the problem of ν_γ. From Theorem 5.3.2 we have $||\phi_{k_i}||_1 \rightarrow \nu_\gamma$. From (5.16) we have $||\phi_1||_1 = \nu_\gamma$. From uniqueness of the optimal solution we have $\phi_1 = \phi_\gamma$. From uniqueness of the optimal it also follows that $\phi_{\gamma_k} \rightarrow \phi_\gamma$. This proves the theorem. □

5.4 An Example

In this section we illustrate the theory developed in the previous sections with an example. Consider the SISO plant,

$$\hat{P}(\lambda) = \lambda - \frac{1}{2}, \tag{5.17}$$

where we are interested in the sensitivity map $\phi := (I - PK)^{-1}$. Using Youla parametrization we get that all achievable transfer functions are given by $\hat{\phi} = (I - \hat{P}\hat{K})^{-1} = 1 - (\lambda - \frac{1}{2})\hat{q}$ where $\hat{q}$ is a stable map. The matrix A and b are given by

$$A = (1, \frac{1}{2}, \frac{1}{2^2}, \ldots), \quad b = 1.$$

It is easy to check that for this problem

$$\mu_\infty := \inf\{||\phi||_2^{\,2} : \phi \in \ell_1 \ and \ A\phi = b\} = 0.75,$$

and

$$m_1 := \inf_{A\phi = b, ||\phi||_2^{\,2} \leq \mu_\infty} ||\phi||_1 = 1.5,$$

with the optimal solution ϕ_2 given by

$$\hat{\phi}_2(\lambda) = \sum_{t=0}^{\infty} \frac{0.75}{2^t} \lambda^t.$$

Performing a standard ℓ_1 optimization [3] we obtain

$$\nu_\infty := inf\{||\phi||_1 : \phi \in \ell_1 \ and \ A\phi = b\} = 1$$

and

$$m_2 := \inf_{A\phi = b, ||\phi||_1 \leq \nu_\infty} ||\phi||_2^{\,2} = 1,$$

with the optimal solution $\phi_1 = 1$. We choose the constraint level to be 0.95. Therefore, $\alpha_\gamma = \frac{2m_1\sqrt{\gamma}}{\gamma - \mu_\infty} + 1 = 15.62$. For this example $n = 1$ and $z_1 = \frac{1}{2}$. L^* the *a priori* bound on the length of the optimal is chosen to satisfy

$$\max_{k=1,\ldots,n} |z_k|^{L^*} \ || \ (A_L^*)^{-l} \ ||_{\infty,1} \ \alpha_\gamma < 1. \tag{5.18}$$

where L is any positive integer such that $L \geq (n-1)$. We choose $L = 0$ and therefore $A_L = 1$ and $|| \ (A_L^*)^{-l} \ ||_{\infty,1} = 1$. We choose $L^* = 4$ which satisfies (5.18). Therefore, the optimal solution ϕ_0 satisfies $\phi_0(i) = 0$ if $i \geq 4$. The problem reduces to the following finite dimensional convex optimization problem:

$$\nu_\gamma = \min_{A_{L^*}\phi = 1, ||\phi||_2^{\,2} \leq 0.95} \{\sum_{k=0}^{3} |\phi(k)| : \phi \in R^4\},$$

where $A_{L^*} = (1, \frac{1}{2}, \frac{1}{2^2}, \frac{1}{2^3})$. We obtain (using Matlab Optimization Toolbox) the optimal solution ϕ_0 to be:

$$\hat{\phi}_0(\lambda) = 0.9732 + 0.0535\lambda.$$

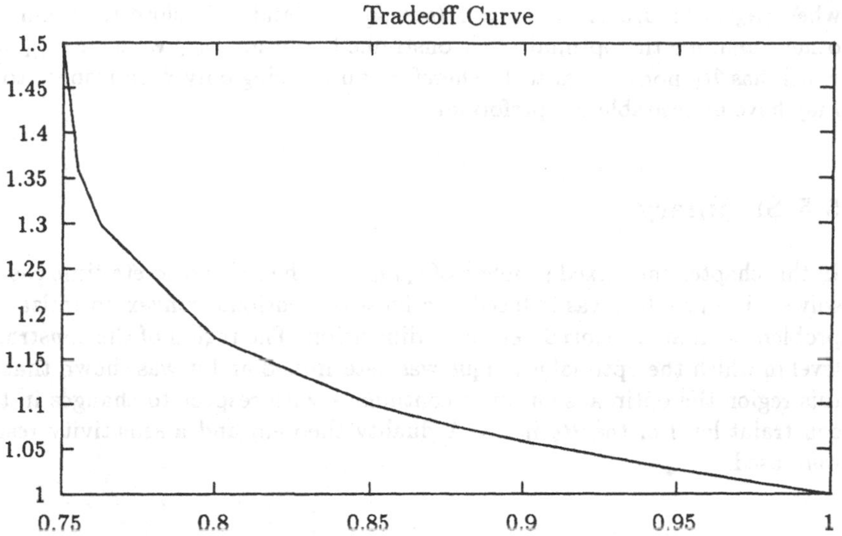

Fig. 5.3. The ℓ_1 and the $\mathcal{H}_2$ norms of the optimal closed loop for various values of γ are plotted. The x axis can be read as the square of $\mathcal{H}_2$ norm or the value of γ. The y axis shows the ℓ_1 norm.

Therefore we have $||\phi_0||_1 = 1.02670$ and $||\phi_0||_2{}^2 \cong 0.95$. The same computation was carried out for various values of the constraint level, $\gamma \in [0.75,\ 1]$. The tradeoff curve between the ℓ_1 and the $\mathcal{H}_2$ norms of the optimal solution is given in Figure 5.3. For all values of γ in the chosen range the square of the $\mathcal{H}_2$ norm of the optimal closed loop was equal to the constraint level γ. Although when the constraint level γ equals 0.75 the optimal closed loop map is an infinite impulse response sequence, the optimal closed loop map has very few nonzero terms in its impulse response even for values of γ very close to 0.75. For example with $\gamma = 0.755$ the optimal closed loop map is given by:

$$\hat{\phi}_{0.755} = 0.7708 + 0.3632\lambda + 0.1596\lambda^2 + 0.0578\lambda^3 + 0.0065\lambda^4.$$

As a final remark, we can use the structure of this example to illustrate that the optimal unconstrained $\mathcal{H}_2$ solution can have $\mathcal{H}_2$ norm much smaller than the $\mathcal{H}_2$ norm of the optimal ℓ_1 (unconstrained) solution. Hence, minimizing only the ℓ_1 norm, which is an upper bound on the $\mathcal{H}_2$ norm, may require substantial sacrifices in terms of $\mathcal{H}_2$ performance. Indeed, instead of the P used in the example before, consider the plant $\hat{P}_a(\lambda) = \lambda - a$ where now a is a zero in the unit disk (i.e., $|a| < 1$) and very close to the unit circle (i.e., $|a| \cong 1$). Then the optimal unconstrained $\mathcal{H}_2$ norm given by

$$(b_a(A_a A_a^*)^{-1} b_a)^{1/2} = (1 - |a|^2)^{1/2}$$

where $A_a = (1, a, a^2, \ldots)$, $b_a = 1$ (see [3] for details) is close to 0. On the other hand, for the optimal ℓ_1 unconstrained solution $\phi_{a,1}$ we have $\phi_{a,1} = 1$ which has $\mathcal{H}_2$ norm equal to 1. Therefore minimizing only with respect to ℓ_1 may have undesirable $\mathcal{H}_2$ performance.

5.5 Summary

In this chapter the mixed problem of $\ell_1/\mathcal{H}_2$ for the SISO discrete time case is solved. The problem was reduced to a finite dimensional convex optimization problem with an *a priori* determined dimension. The region of the constraint level in which the optimal is unique was determined and it was shown that in this region the optimal solution is continuous with respect to changes in the constraint level of the $\mathcal{H}_2$ norm. A duality theorem and a sensitivity result were used.

6. A Composite Performance Measure

This chapter studies a "mixed" objective problem of minimizing a composite measure of the ℓ_1, $\mathcal{H}_2$, and ℓ_∞ norms together with the ℓ_∞ norm of the step response of the closed loop . This performance index can be used to generate Pareto optimal solutions with respect to the individual measures. The problem is analysed for the discrete time, single-input single-output (SISO), linear time invariant systems. It is shown via the Lagrange duality theory that the problem can be reduced to a convex optimization problem with *a priori* known dimension. In addition, continuity of the unique optimal solution with respect to changes in the coefficients of the linear combination is established.

6.1 Problem Formulation

Let w_1 be the unit step input i.e., $w_1 = (1, 1, \ldots)$. The problem of interest can be stated as:
Given $c_1 > 0$, $c_2 > 0$, $c_3 > 0$, and $c_4 > 0$ obtain a solution to the following mixed objective problem:

$$\begin{aligned} \nu :=& \inf_{\phi \text{ Achievable}} \{c_1||\phi||_1 + c_2||\phi||_2^2 + c_3||\phi * w_1||_\infty + c_4||\phi||_\infty \} \\ =& \inf_{\phi\in\ell_1,\ A\phi=b} \{c_1||\phi||_1 + c_2||\phi||_2^2 + c_3||\phi * w_1||_\infty + c_4||\phi||_\infty\}. \end{aligned} \quad (6.1)$$

The assumptions made on the plant are the same assumptions that were made in Chapter 5. The definitions of achievability, the matrix A and the vector b are as given in Chapter 5. We define $f : \ell_1 \to R$ by,

$$f(\phi) := c_1||\phi||_1 + c_2||\phi||_2^2 + c_3||\phi * w_1||_\infty + c_4||\phi||_\infty,$$

which is the objective functional in the optimization given by (6.1).

In the following sections we will study the existence, structure and computation of the optimal solution. Before we initiate our study towards these goals it is worthwhile to point out certain connections between the cost under consideration and the notion of Pareto optimality.

6.1.1 Relation to Pareto Optimality

The notion of Pareto optimality can be stated as follows (see for example, [22]). Given a set of m nonnegative functionals $\overline{f}_i$, $i = 1, \ldots, m$ on a normed linear space X, a point $x_0 \in X$ is Pareto optimal with respect to the vector valued criterion $\overline{f} := (\overline{f}_1, \ldots, \overline{f}_m)$ if there does not exist any $x \in X$ such that

$\overline{f}_i(x) \leq \overline{f}_i(x_0)$ for all $i \in \{1, \ldots, m\}$ and $\overline{f}_i(x) < \overline{f}_i(x_0)$ for some $i \in \{1, \ldots, m\}$.

Under certain conditions the set of all Pareto optimal solutions can be generated by solving a minimization of weighted sum of the functionals as the following theorem indicates.

Theorem 6.1.1. *[23] Let X be a normed linear space and each nonnegative functional $\overline{f}_i$ be convex. Also let*

$$S_m := \{c \in R^m : c_i \geq 0, \sum_{i=1}^{m} c_i = 1\},$$

and for each $c \in R^m$ consider the following scalar valued optimization:

$$\inf_{x \in X} \sum_{i=1}^{m} c_i \overline{f}_i(x).$$

If $x_0 \in X$ is Pareto optimal with respect to the vector valued criterion $\overline{f}(x)$, then there exists some $c \in S_m$ such that x_0 solves the above minimization. Conversely, given $c \in S_m$, if the above minimization has at most one solution x_0 then x_0 is Pareto optimal with respect to $\overline{f}(x)$. □

In the next section we show that there is a unique solution ϕ_0 to Problem (6.1). Furthermore, since u is assumed to be a scalar, there is a unique optimal $q \in \ell_1$. Hence, in view of the aforementioned theorem we have that if we restrict our attention to parameters c_1, c_2, c_3 and c_4 such that $(c_1, c_2, c_3, c_4) \in \Sigma_4 := \{(c_1, c_2, c_3, c_4) : c_1 + c_2 + c_3 + c_4 = 1,\ c_1, c_2, c_3, c_4 > 0\}$, we will produce a set of Pareto optimal solutions with respect to the vector valued function

$$\overline{f}(q) := (||h - u * q||_1, ||h - u * q||_2{}^2, ||(h - u * q) * w_1||_\infty, ||h - u * q||_\infty) =: (\overline{f}_1(q), \overline{f}_2(q), \overline{f}_3(q), \overline{f}_4(q)).$$

where $q \in \ell_1$. Thus, if ϕ_0 is the optimal solution for Problem (6.1) with a corresponding q_0 for some given $(c_1, c_2, c_3, c_4) \in \Sigma_4$, then there does not exist a preferable alternative ϕ with $\phi = h - u * q$ for some $q \in \ell_1$ such that

$\overline{f}_i(q) \leq \overline{f}_i(q_0)$ for all $i \in \{1, \ldots, 4\}$ and $\overline{f}_i(q) < \overline{f}_i(q_0)$ for some $i \in \{1, \ldots, 4\}$.

As a final note we mention that if (c_1, c_2, c_3, c_4) do not satisfy $c_1 + c_2 + c_3 + c_4 = 1$ then we can define a new set of parameters $\bar{c}_1$, $\bar{c}_2$, $\bar{c}_3$ and $\bar{c}_4$ by $\bar{c}_1 = \frac{c_1}{c_1+c_2+c_3+c_4}$, $\bar{c}_2 = \frac{c_2}{c_1+c_2+c_3+c_4}$, $\bar{c}_3 = \frac{c_3}{c_1+c_2+c_3+c_4}$ and $\bar{c}_4 = \frac{c_4}{c_1+c_2+c_3+c_4}$ with $\bar{c}_1 + \bar{c}_2 + \bar{c}_3 + \bar{c}_4 = 1$. These new parameters would yield the same optimal solution as with (c_1, c_2, c_3, c_4).

6.2 Properties of the Optimal Solution

In the first part of this section we show that Problem (6.1) always has a solution. In the second part we show that any solution to Problem (6.1) is a finite impulse response sequence and in the third we give an *a priori* bound on the length.

6.2.1 Existence of a Solution

Here we show that a solution to (6.1) always exists.

Theorem 6.2.1. *There exists $\phi_0 \in \Phi$ such that*

$$f(\phi_0) = \inf_{\phi \in \Phi} \{c_1||\phi||_1 + c_2||\phi||_2^2 + c_3||\phi * w_1||_\infty + c_4||\phi||_\infty\},$$

where $\Phi := \{\phi \in \ell_1 : A\phi = b\}$. Therefore the infimum in (6.1) is a minimum.

Proof. We denote the feasible set of our problem by $\Phi := \{\phi \in \ell_1 : A\phi = b\}$. Let

$$B := \{\phi \in \ell_1 : c_1||\phi||_1 + c_2||\phi||_2^2 + c_3||\phi * w_1||_\infty + c_4||\phi||_\infty \leq \nu + 1\}.$$

It is clear that

$$\nu = \inf_{\phi \in \Phi \cap B} \{c_1||\phi||_1 + c_2||\phi||_2^2 + c_3||\phi * w_1||_\infty + c_4||\phi||_\infty\}.$$

Therefore given $i > 0$ there exists $\phi_i \in \Phi \cap B$ such that

$$c_1||\phi_i||_1 + c_2||\phi_i||_2^2 + c_3||\phi_i * w_1||_\infty + c_4||\phi_i||_\infty \leq \nu + \frac{1}{i}.$$

Let

$$\overline{B} := \{\phi \in \ell_1 : c_1||\phi||_1 \leq \nu + 1\}.$$

$\overline{B}$ is a bounded set in $\ell_1 = c_0^*$. It follows from the Banach-Alaoglu result (see Theorem 2.3.1) that $\overline{B}$ is $W(c_0^*, c_0)$ compact. Using the fact that c_0 is separable and that $\{\phi_i\}$ is a sequence in $\overline{B}$ we know that there exists a subsequence $\{\phi_{i_k}\}$ of $\{\phi_i\}$ and $\phi_0 \in \overline{B}$ such that $\phi_{i_k} \to \phi_0$ in the $W(c_0^*, c_0)$ sense, that is for all v in c_0

$$< v, \phi_{i_k} > \to < v, \phi_0 > \text{ as } k \to \infty. \tag{6.2}$$

Let the jth row of A be denoted by a_j and the jth element of b be given by b_j . Then as $a_j \in c_0$ we have,

$$< a_j, \phi_{i_k} > \to < a_j, \phi_0 > \text{ as } k \to \infty \text{ for all } j = 1, 2, \ldots, n. \tag{6.3}$$

As $A(\phi_{i_k}) = b$ we have $< a_j, \phi_{i_k} >= b_j$ for all k and for all j which implies $< a_j, \phi_0 >= b_j$ for all j. Therefore we have $A(\phi_0) = b$ from which it follows that $\phi_0 \in \Phi$. This gives us $c_1||\phi_0||_1 + c_2||\phi_0||_2^2 + c_3||\phi_0 * w_1||_\infty + c_4||\phi_0||_\infty \geq \nu$.

From (6.2) we can deduce that for all t, $\phi_{i_k}(t)$ converges to $\phi_0(t)$. An easy consequence of this is we have for all N as k tends to ∞, $\sum_{t=0}^{N}\{c_1|\phi_{i_k}(t)| + c_2(\phi_{i_k}(t))^2\} + c_3 \max_{0\le t\le N} |(\phi_{i_k} * w_1)(t)| + c_4 \max_{0\le t\le N} |\phi_{i_k}(t)|$ converges to $\sum_{t=0}^{N}\{c_1|\phi_0(t)| + c_2(\phi_0(t))^2\} + c_3 \max_{t\le N} |(\phi_0 * w_1)(t)| + c_4 \max_{0\le t\le N} |\phi_0(t)|$. As $f(\phi_{i_k}) \le \nu + \frac{1}{i_k}$ we have $\sum_{t=0}^{N}\{c_1|\phi_{i_k}(t)| + c_2(\phi_{i_k}(t))^2\} + c_3 \max_{t\le N} |(\phi_{i_k} * w_1)(t)| + c_4 \max_{0\le t\le N} |\phi_{i_k}(t)| \le \nu + \frac{1}{i_k}$. Letting $k \to \infty$ we have that for all N

$$\sum_{t=0}^{N}\{c_1|\phi_0(t)| + c_2(\phi_0(t))^2\} + c_3 \max_{t\le N} |(\phi_0 * w_1)(t)| + c_4 \max_{0\le t\le N} |\phi_0(t)| \le \nu.$$

Letting $N \to \infty$ in the above inequality we conclude that $c_1\|\phi_0\|_1 + c_2\|\phi_0\|_2{}^2 + c_3\|\phi_0 * w_1\|_\infty + c_4\|\phi_0\|_\infty \le \nu$. This proves the theorem. □

6.2.2 Structure of Optimal Solutions

In this subsection we use a Lagrange duality result to show that every optimal solution is of finite length.

Lemma 6.2.1.

$$\nu = \max_{y\in R^n} \inf_{\phi\in\ell_1} \{f(\phi) + < b - A\phi, y >\}. \tag{6.4}$$

Proof. We will apply Theorem 3.3.2 (Kuhn-Tucker-Lagrange duality theorem) to get the result. Let X, Ω, Y, Z in Theorem 3.3.2 correspond to ℓ_1, ℓ_1, R^n, R respectively. Let $\gamma := \nu + 1$, $g(\phi) := f(\phi) - \gamma$ and $H(\phi) := b - A\phi$. With this notation we have $Z^* = R$. A has full range which implies $0 \in int[range(H)]$. From Theorem 6.2.1 we know that there exists ϕ_0 such that $g(\phi_0) = f(\phi_0) - \gamma = -1 < 0$ and $H(\phi_0) = 0$. Therefore all the conditions of Theorem 3.3.2 are satisfied. From Theorem 3.3.2 we have

$$\nu = \max_{z\ge 0, y\in R^n} \inf_{\phi\in\ell_1} \{f(\phi) + < g(\phi), z > + < b - A\phi, y >\}.$$

Let $z_0 \in R$, $y_0 \in R^n$ be a maximizing solution to the right hand side of the above equation. ϕ_0 being the solution of the primal we have from (3.30) that $< g(\phi_0), z_0 > + < H(\phi_0), y_0 >= 0$ which implies that $< g(\phi_0), z_0 >= 0$. As $g(\phi_0) \ne 0$ we conclude that $z_0 = 0$. This proves the theorem.□

The following theorem shows that the solution to (6.1) is unique and that it is a finite impulse response sequence.

Theorem 6.2.2. *Define* $\mathcal{T} := \{\phi \in \ell_1$ *: there exists* L^* *with* $\phi(i) = 0$ *if* $i \ge L^*\}$. *The following is true:*

$$\nu = \max_{y\in R^n} \inf_{\phi\in\mathcal{T}} \{f(\phi) - < \phi, v > + < b, y >\}, \tag{6.5}$$

where $v(i) := (A^*y)(i)$. *Also, the solution to the primal (6.1) is unique and the solution belongs to* $\mathcal{T}$.

Proof. Let $y_0 \in R^n$ be the solution to the right hand side of (6.5). Define $v_0 := (A^* y_0)(i)$ and let

$$J(\phi) := f(\phi) - < \phi, v_0 > + < b, y_0 > .$$

It is immediate that ν is equal to

$$\inf_{\phi \in \ell_1} \sum_{i=0}^{\infty} \{c_1|\phi(i)| + c_2(\phi(i))^2 - \phi(i)v_0(i)\} + c_3 \sup_i |(\phi * w_1)(i)| + c_4 \sup_i |\phi(i)| + < b, y_0 > .$$

As v_0 is in ℓ_1 we know that there exists L^* such that $v_0(i)$ satisfies $|v_0(i)| < c_1$ if $i \geq L^*$. We now show that for ϕ to be optimal it is necessary that $\phi(i) = 0$ for all $i \geq L^*$. Indeed, if $\phi(i) \neq 0$ while $|v_0(i)| < c_1$ for some i, note that,

$$c_1|\phi(i)| + c_2(\phi(i))^2 - \phi(i)v_0(i) > c_1|\phi_1(i)| + c_2(\phi_1(i))^2 - \phi_1(i)v_0(i),$$

for any $\phi_1 \in \ell_1$ such that $\phi_1(i) = 0$. Moreover, if ϕ_1 is such that $\phi_1(j) = \phi(j)$ whenever $j < L^*$ and $\phi_1(j) = 0$ whenever $j \geq L^*$ it follows that

$$\sup_i |\sum_{j=0}^{i} \phi(j)| \geq \sup_i |\sum_{j=0}^{i} \phi_1(j)| \text{ and } \sup_i |\phi(i)| \geq \sup_i |\phi_1(i)|$$

or equivalently,

$$||\phi * w_1||_\infty \geq ||\phi_1 * w_1||_\infty \text{ and } ||\phi||_\infty \geq ||\phi_1||_\infty .$$

Hence, we have that $J(\phi) > J(\phi_1)$ which proves our claim. In Theorem 6.2.1 we showed that there exists a solution ϕ_0 to the primal (6.1). From Theorem 3.3.2 we know that ϕ_0 is a solution to $\inf_{\phi \in \ell_1} J(\phi)$. As $J(\phi)$ is strictly convex in ϕ we conclude that the solution to the primal (6.1) is unique. From the previous discussion it follows that the solution to the primal (6.1), ϕ_0 is in $\mathcal{T}$. This proves the theorem. □

6.2.3 An A priori Bound on the Length of any Optimal Solution

In this subsection we give an *a priori* bound on the length of the solution to (6.1). First we establish the following two lemmas.

Lemma 6.2.2. *Let ϕ_0 be a solution of the primal (6.1). Let y_0 represent a corresponding dual solution as obtained in (6.5). Let $v_0 := A^* y_0$ then, $||v_0||_\infty \leq \alpha$ where $\alpha = c_1 + c_3 + c_4 + 2\frac{c_2}{c_1} f(h)$.*

Proof. From the proof of Theorem 6.2.2 it is clear that ϕ_0 should be such that it minimizes $J(\phi)$ given by

$$\sum_{i=0}^{L^*} \{c_1|\phi(i)| + c_2(\phi(i))^2 - \phi(i)v_0(i)\} + c_3 \max_{i \leq L^*} |\sum_{j=0}^{i} \phi(j)| + c_4 \max_{i \leq L^*} |\phi(i)|,$$

where L^* is such that $|v_0(i)| < c_1$ if $i \geq L^*$.

Let i be any integer such that $i \leq L^*$. Consider perturbation ϕ of ϕ_0 given as $\phi(i) = \phi_0(i) + \epsilon$ and $\phi(j) = \phi_0(j)$ for $j \neq i$. Then, for all ϵ, it can be shown that

$$\max_{0 \leq t \leq L^*} |\sum_{j=0}^{t} \phi(j)| - \max_{0 \leq t \leq L^*} |\sum_{j=0}^{t} \phi_0(j)| \leq |\epsilon| \tag{6.6}$$

and

$$\max_{0 \leq t \leq L^*} |\phi(t)| - \max_{0 \leq t \leq L^*} |\phi_0(t)| \leq |\epsilon|. \tag{6.7}$$

Indeed, assume that

$$\max_{0 \leq t \leq L^*} |\sum_{j=0}^{t} \phi_0(j)| = |\sum_{j=0}^{N} \phi_0(j)| \text{ for some } N \leq L^*.$$

For the given ϵ let

$$\max_{0 \leq t \leq L^*} |\sum_{j=0}^{t} \phi(j)| = |\sum_{j=0}^{M} \phi(j)| \text{ for some } M \leq L^*. \tag{6.8}$$

If $M \leq i$ then

$$\begin{aligned}\max_{0 \leq t \leq L^*} |\sum_{j=0}^{t} \phi(j)| - \max_{0 \leq t \leq L^*} |\sum_{j=0}^{t} \phi_0(j)| &= |\phi_0(0) + \ldots + \phi_0(M)| \\ &\quad - |\phi_0(0) + \ldots + \phi_0(N)| \\ &\leq 0 \leq |\epsilon|,\end{aligned}$$

and if $M > i$ then

$$\begin{aligned}\max_{0 \leq t \leq L^*} |\sum_{j=0}^{t} \phi(j)| - \max_{0 \leq t \leq L^*} |\sum_{j=0}^{t} \phi_0(j)| &= |\phi_0(0) + \ldots + (\phi_0(i) + \epsilon) + \ldots \\ &\quad + \phi_0(M)| - |\phi_0(0) + \ldots + \phi_0(N)| \\ &\leq |\phi_0(0) + \ldots + \phi_0(M)| + |\epsilon| \\ &\quad - |\phi_0(0) + \ldots + \phi_0(N)| \\ &\leq |\epsilon|.\end{aligned}$$

(6.7) can be proved easily. It follows from (6.6) and (6.7) that

$$\begin{aligned}J(\phi) - J(\phi_0) &= c_1(|\phi_0(i) + \epsilon| - |\phi_0(i)|) + c_2(\epsilon^2 + 2\epsilon\phi_0(i)) \\ &\quad + c_3 \max_{0 \leq t \leq L^*} |\sum_{j=0}^{t} \phi(j)| - c_3 \max_{0 \leq t \leq L^*} |\sum_{j=0}^{t} \phi_0(j)|) \\ &\quad + c_4(\max_{0 \leq t \leq L^*} |\phi(t)| - \max_{0 \leq t \leq L^*} |\phi_0(t)|) - \epsilon v_0(i) \\ &\leq c_1|\epsilon| + c_2(\epsilon^2 + 2\epsilon\phi_0(i)) + c_3|\epsilon| + c_4|\epsilon| - \epsilon v_0(i).\end{aligned}$$

As ϕ_0 is the unique minimum we have that $J(\phi) - J(\phi_0) > 0$ and therefore it follows that

$$c_1|\epsilon| + c_2(\epsilon^2 + 2\epsilon\phi_0(i)) + c_3|\epsilon| + c_4|\epsilon| - \epsilon v_0(i) > 0 \text{ for all } \epsilon.$$

Dividing both sides of the above inequality by $|\epsilon|$ we get

$$c_1 + c_3 + c_4 + c_2|\epsilon| + 2c_2\frac{\epsilon}{|\epsilon|}\phi_0(i) - \frac{\epsilon}{|\epsilon|}v_0(i) > 0 \text{ for all } \epsilon.$$

Letting $\epsilon \to 0^+$ and $\epsilon \to 0^-$ in the above inequality we have

$$v_0(i) \leq c_1 + c_3 + c_4 + 2c_2|\phi_0(i)|$$

and

$$-v_0(i) \leq c_1 + c_3 + c_4 - 2c_2|\phi_0(i)| \leq c_1 + c_3 + c_4 + 2c_2|\phi_0(i)|,$$

respectively. This implies that

$$|v_0(i)| \leq c_1 + c_3 + c_4 + 2c_2|\phi_0(i)| \leq c_1 + c_3 + c_4 + 2c_2||\phi_0||_1$$

As this holds for any $i \leq L^*$ we have

$$||v_0||_\infty \leq c_1 + c_3 + c_4 + 2\frac{c_2}{c_1}f(\phi_0) \leq c_1 + c_3 + c_4 + 2\frac{c_2}{c_1}f(h),$$

where we have used that $2c_2||\phi_0||_1 \leq 2\frac{c_2}{c_1}f(\phi_0) \leq c_1 + c_3 + c_4 + 2\frac{c_2}{c_1}f(h)$, and, $f(\phi_0) \leq f(h)$ since h is feasible ($q = 0$) but not necessarily optimal. This proves the lemma.□

Lemma 6.2.3. *If $y \in R^n$ is such that $||A^*y||_\infty \leq \alpha$ then there exists a positive integer L^* independent of y such that $|(A^*y)(i)| < c_1$ for all $i \geq L^*$.*

Proof. Define

$$A_L^* = \begin{pmatrix} 1 & 1 & 1 & \dots & 1 \\ z_1 & z_2 & z_3 & \dots & z_n \\ \vdots & \vdots & \vdots & \vdots & \vdots \\ z_1^L & z_2^L & z_3^L & \dots & z_n^L \end{pmatrix},$$

$A_L^* : R^n \to R^{L+1}$. With this definition we have $A_\infty^* = A^*$. Let $y \in R^n$ be such that $||A^*y||_\infty \leq \alpha$. Choose any L such that $L \geq (n-1)$. As $z_i, i = 1, \dots, n$ are distinct A_L^* has full column rank. A_L^* can be regarded as a linear map taking $(R^n, ||.||_1) \to (R^{L+1}, ||.||_\infty)$. As A_L^* has full column rank we can define the left inverse of A_L^*, $(A_L^*)^{-l}$ which takes $(R^{L+1}, ||.||_\infty) \to (R^n, ||.||_1)$. Let the induced norm of $(A_L^*)^{-l}$ be given by $||\ (A_L^*)^{-l}\ ||_{\infty,1}$. $y \in R^n$ is such that $||A^*y||_\infty \leq \alpha$ and therefore $||A_L^*y||_\infty \leq \alpha$. It follows that,

$$||y||_1 \leq ||\ (A_L^*)^{-l}\ ||_{\infty,1}\ ||A_L^*y||_\infty \leq ||\ (A_L^*)^{-l}\ ||_{\infty,1}\ \alpha. \tag{6.9}$$

Choose L^* such that

$$\max_{k=1,\dots,n} |z_k|^{L^*}\ ||\ (A_L^*)^{-l}\ ||_{\infty,1}\ \alpha < c_1. \tag{6.10}$$

There always exists such an L^* because $|z_k| < 1$ for all $k = 1, \ldots, n$. Note that L^* does not depend on y. For any $i \geq L^*$ we have

$$\begin{aligned}
|(A^*y)(i)| = |\sum_{k=1}^{k=n} z_k^i y(k)| &\leq \max_{k=1,\ldots,n} |z_k|^i \|y\|_1 \\
&\leq \max_{k=1,\ldots,n} |z_k|^i \ \| (A_L^*)^{-l} \|_{\infty,1} \, \alpha \\
&\leq \max_{k=1,\ldots,n} |z_k|^{L^*} \ \| (A_L^*)^{-l} \|_{\infty,1} \, \alpha.
\end{aligned}$$

The second inequality follows from 6.9. From 6.10 we have $|(A^*y)(i)| < c_1$ if $i \geq L^*$. This proves the lemma. □

We now summarize the main result of this subsection

Theorem 6.2.3. *The unique solution ϕ_0 of the primal (6.1) is such that $\phi(i) = 0$ if $i \geq L^*$ where L^* given in Lemma 6.2.3 can be determined a priori.*

Proof. Let y_0 be the dual solution to (6.1) and let $v_0 := A^*y_0$. From Lemma 6.2.2 we know that $\|v_0\|_\infty \leq \alpha$ where $\alpha = c_1 + c_3 + c_4 + 2\frac{c_2}{c_1} f(h)$. Applying Lemma 6.2.3 we conclude that there exists L^* (which can be determined a priori) such that $|v_0(i)| < c_1$ if $i \geq L^*$. Therefore, $\phi_0(i) = 0$ if $i \geq L^*$. This proves the theorem. □

The above theorem shows that the Problem (6.1) is a finite dimensional convex minimization problem. Such problems can be solved efficiently using standard numerical methods.

At this point we would like to make a few remarks. It should be clear that the uniqueness property of the optimal solution is due the the non-zero coefficient c_2. This makes the problem strictly convex. The finite impulse response property of the optimal solution is due to the nonzero c_1. Also, it should be noted that in the case where c_3 and/or c_4 are allowed to be zero, all of the previous results apply by setting respectively c_3 and/or c_4 to zero in the appropriate expressions for the upper bounds.

6.3 An Example

In this section we illustrate the theory developed in the previous sections with an example taken from [14] and also considered in Chapter 5. Consider the SISO plant,

$$\hat{P}(\lambda) = \lambda - \frac{1}{2}, \tag{6.11}$$

where we are interested in the sensitivity map $\phi := (I - PK)^{-1}$. Using Youla parametrization we get that all achievable transfer functions are given by

$\hat{\phi} = (I - \hat{P}\hat{K})^{-1} = 1 - (\lambda - \frac{1}{2})\hat{q}$ where $\hat{q}$ is a stable map. Therefore, $h = 1$ and $u = \lambda - \frac{1}{2}$. The matrix A and b are given by

$$A = (1, \frac{1}{2}, \frac{1}{2^2}, \ldots), \quad b = 1.$$

We consider the case where $c_1 = 1, c_2 = 1, c_3 = 1$ and $c_4 = 1$. Therefore, $\alpha = c_1 + c_3 + c_4 + 2\frac{c_2}{c_1}f(h)) = 11$. For this example $n = 1$ and $z_1 = \frac{1}{2}$. L^* the *a priori* bound on the length of the optimal is chosen to satisfy

$$\max_{k=1,\ldots,n} |z_k|^{L^*} \| (A_L^*)^{-l} \|_{\infty,1} \alpha < c_1, \tag{6.12}$$

where L is any positive integer such that $L \geq (n-1)$. We choose $L = 0$ and therefore $A_L = 1$ and $\| (A_L^*)^{-l} \|_{\infty,1} = 1$. We choose $L^* = 4$ which satisfies (6.12). Therefore, the optimal solution ϕ_0 satisfies $\phi_0(i) = 0$ if $i \geq 4$. The problem reduces to the following finite dimensional convex optimization problem:

$$\nu = \min_{A_{L^*}\phi = 1} \{\sum_{k=0}^{3}(|\phi(k)| + (\phi(k))^2) + \max_{0 \leq k \leq 3} |(\phi * w_1)(k)| + \max_{0 \leq k \leq 3} |\phi(k)| : \phi \in R^4\},$$

where $A_{L^*} = (1, \frac{1}{2}, \frac{1}{4}, \frac{1}{8})$. We obtain (using Matlab Optimization Toolbox) the optimal solution ϕ_0 to be:

$$\hat{\phi}_0(\lambda) = 0.9 + 0.2\lambda.$$

6.4 Continuity of the Optimal Solution

In this section we show that the optimal is continuous with respect to changes in the parameters c_1, c_2, c_3 and c_4. First, we prove the following lemma:

Lemma 6.4.1. *Let $\{f_k\}$ be a sequence of functions which map R^m to R. If f_k converges uniformly to a function f on a set $S \subset R^m$ then*

$$\lim_{k \to \infty} \min_{x \in S} f_k(x) = \min_{x \in S} f(x),$$

provided that the minima exist.

Proof. Let $\min_{x \in S} f(x) = f(x_0)$ for some $x_0 \in S$. Given $\epsilon > 0$ we know from convergence of the sequence $\{f_k\}$ to f that there exists an integer K such that if $k > K$ then

$$|f_k(x_0) - f(x_0)| < \epsilon,$$

$$\Rightarrow f_k(x_0) < \epsilon + f(x_0),$$

$$\Rightarrow \min_{x\in S} f_k(x) < \epsilon + f(x_0),$$

$$\Rightarrow \lim_{k\to\infty} \min_{x\in S} f_k(x) < \epsilon + f(x_0).$$

As ϵ is arbitrary we have $\lim_{k\to\infty} \min_{x\in S} f_k(x) \leq f(x_0)$. Now we prove the other inequality. Given $\epsilon > 0$ we know that there exists an integer K such that if $k > K$ then

$$|f_k(x) - f(x)| < \epsilon \text{ for any } x \in S$$

$$\Rightarrow f_k(x) > f(x) - \epsilon \geq f(x_0) - \epsilon \text{ for any } x \in S$$

$$\Rightarrow \min_{x\in S} f_k(x) > f(x_0) - \epsilon$$

$$\Rightarrow \lim_{k\to\infty} \min_{x\in S} f_k(x) > f(x_0) - \epsilon.$$

As ϵ is arbitrary we have $\lim_{k\to\infty} \min_{x\in S} f_k(x) \geq f(x_0)$. This proves the lemma □

Theorem 6.4.1. *Let* $c_1^k \in [a_1, b_1]$, $c_2^k \in [a_2, b_2]$, $c_3^k \in [a_3, b_3]$ *and* $c_4^k \in [a_4, b_4]$ *where* $a_1 > 0$, $a_2 > 0$, $a_3 > 0$, $a_4 > 0$. *Let* ϕ_k *be the unique solution to the problem*

$$\nu_k := \min_{A\phi=b} c_1^k||\phi||_1 + c_2^k||\phi||_2^2 + c_3^k||\phi * w_1||_\infty + c_4^k||\phi||_\infty, \tag{6.13}$$

and let ϕ_0 *be the solution to the problem*

$$\nu := \min_{A\phi=b} c_1||\phi||_1 + c_2||\phi||_2^2 + c_3||\phi * w_1||_\infty + c_4||\phi||_\infty, \tag{6.14}$$

If $c_1^k \to c_1$, $c_2^k \to c_2$, $c_3^k \to c_3$ *and* $c_4^k \to c_4$ *then* $\phi_k \to \phi_0$.

Proof. We prove this theorem in three parts; first we show that we can restrict the proof to a finite dimensional space, second we show that $\nu_k \to \nu$ and finally we show that $\phi_k \to \phi_0$. Let y_k represent the dual solution of (6.13) and let $v_k := A^* y_k$. Let $f_k(\phi) := c_1^k||\phi||_1 + c_2^k||\phi||_2^2 + c_3^k||\phi_k * w_1||_\infty + c_4^k||\phi||_\infty$ and $f(\phi) := c_1||\phi||_1 + c_2||\phi||_2^2 + c_3||\phi * w_1||_\infty + c_4||\phi||_\infty$. Let α_k the upper bound on $||v_k||_\infty$ be as given by Lemma 6.2.2. Therefore,

$$\begin{aligned}\alpha_k &= c_1^k + c_3^k + c_4^k + 2\tfrac{c_2^k}{c_1^k} f_k(h)\\ &\leq b_1 + b_3 + b_4 + 2b_2(||h||_1 + \tfrac{b_2}{a_1}||h||_2^2 + \tfrac{b_3}{a_1}||h * w_1||_\infty + \tfrac{b_4}{a_1}||h||_\infty).\end{aligned}$$

Let this bound be denoted by d. Choose L^* such that

$$\max_{i=1,\ldots,n} |z_i|^{L^*} \;||\, (A_L^*)^{-l} \,||_{\infty,1}\, d < a_1.$$

where L is such that $L \geq (n-1)$. Therefore, it follows that

$$\max_{i=1,\ldots,n} |z_i|^{L^*} \;||\, (A_L^*)^{-l} \,||_{\infty,1}\, \alpha_k < c_1^k.$$

for all k. From arguments similar to that of Lemma 6.2.3 and Theorem 6.2.3 it follows that $\phi_k(i) = 0$ if $i \geq L^*$ for all k. Therefore we can assume that $\phi_k \in R^{L^*}$.

Now, we prove that $\nu_k \to \nu$. Let ϕ_1 be the solution of the problem

$$\nu_1 := \min_{A\phi=b} b_1||\phi||_1 + b_2||\phi||_2{}^2 + b_3||\phi * w_1||_\infty + b_4||\phi||_\infty.$$

As $c_1^k \leq b_1$, $c_2^k \leq b_2$, $c_3^k \leq b_3$ and $c_4^k \leq b_4$ we have that $\nu_k \leq \nu_1$ for all k. Therefore, for any k we have $c_1^k||\phi_k||_1 + c_2^k||\phi_k||_2{}^2 + c_3^k||\phi_k * w_1||_\infty + c_4^k||\phi_k||_\infty \leq \nu_1$ which implies $||\phi_k||_1 \leq \frac{\nu_1}{c_1^k} \leq \frac{\nu_1}{a_1}$, $||\phi_k||_2{}^2 \leq \frac{\nu_1}{c_2^k} \leq \frac{\nu_1}{a_2}$, $||\phi_k * w_1||_\infty \leq \frac{\nu_1}{c_3^k} \leq \frac{\nu_1}{a_3}$ and $||\phi_k||_\infty \leq \frac{\nu_1}{a_4}$. Let $S := \{\phi \in R^{L*} : A\phi = b, ||\phi||_1 \leq \frac{\nu_1}{a_1}, ||\phi||_2{}^2 \leq \frac{\nu_1}{a_2}, ||\phi * w_1||_\infty \leq \frac{\nu_1}{a_3}, ||\phi||_\infty \leq \frac{\nu_1}{a_4}\}$. Then it is clear that

$$\nu_k := \min_{\phi \in S} c_1^k||\phi||_1 + c_2^k||\phi||_2{}^2 + c_3^k||\phi * w_1||_\infty + c_4^k||\phi||_\infty.$$

We now prove that f_k converges to f uniformly on S. Given $\epsilon > 0$ choose K such that if $k > K$ then $|c_1^k - c_1| < \frac{\epsilon a_1}{4\nu_1}$, $|c_2^k - c_2| < \frac{\epsilon a_2}{4\nu_1}$, $|c_3^k - c_3| < \frac{\epsilon a_3}{4\nu_1}$ and $|c_4^k - c_4| < \frac{\epsilon a_4}{4\nu_1}$. Then for any $\phi \in S$ we have $|f_k(\phi) - f(\phi)| = |(c_1^k - c_1)||\phi||_1 + (c_2^k - c_2)||\phi||_2{}^2 + (c_3^k - c_3)||\phi * w_1||_\infty + (c_4^k - c_4)||\phi||_\infty|$ and thus

$$|f_k(\phi) - f(\phi)| \leq |c_1^k - c_1|\frac{\nu_1}{a_1} + |c_2^k - c_2|\frac{\nu_1}{a_2} + |c_3^k - c_3|\frac{\nu_1}{a_3} + |c_4^k - c_4|\frac{\nu_1}{a_4} < \epsilon.$$

Therefore, it follows that f_k converges uniformly to f on S. From Lemma 6.4.1 it follows that $\nu_k \to \nu$.

We now prove that $\phi_k \to \phi_0$. Let $B := \{\phi \in R^{L*} : ||\phi||_1 \leq \frac{\nu_1}{a_1}\}$ then we know that $\phi_k \in B$ which is compact in $(R^{L*}, ||.||_1)$. Therefore there exists a subsequence ϕ_{k_i} of ϕ_k and $\overline{\phi} \in R^{L^*}$ such that $\phi_{k_i} \to \overline{\phi}$.

As $c_1^k \to c_1$, $c_2^k \to c_2$, $c_3^k \to c_3$, $c_4^k \to c_4$, and $\phi_{k_i} \to \overline{\phi}$ we have that $f_{k_i}(\phi_{k_i}) \to f(\overline{\phi})$. As ν_k converges to ν it follows that $f_{k_i}(\phi_{k_i}) \to f(\phi_0)$ (note that $\nu_{k_i} = f_{k_i}(\phi_{k_i})$ and $\nu = f(\phi_0)$) and therefore $f(\overline{\phi}) = f(\phi_0)$. As $A\phi_{k_i} = b$ for all i we have that $A\overline{\phi} = b$. From uniqueness of the solution of (6.14) it follows that $\overline{\phi} = \phi_0$. Therefore we have established that $\phi_{k_i} \to \phi_0$. From uniqueness of the solution of (6.14) it also follows that $\phi_k \to \phi_0$. This proves the theorem. □

6.5 Summary

In this chapter we considered a mixed objective problem of minimizing a given linear combination of the ℓ_1 norm, the square of the $\mathcal{H}_2$ norm, and the ℓ_∞ norms of the step and pulse responses respectively of the closed loop. Employing the Khun-Tucker-Lagrange duality theorem it was shown that this problem is equivalent to a finite dimensional convex optimization problem with an *a priori* known dimension. The solution is unique and represents a

Pareto optimal point with respect to the individual measures involved. It was also shown that the optimal solution is continuous with respect to changes in the coefficients of the composite measure.

7. MIMO Design: The Square Case

This chapter and the next chapter explore the interplay of the ℓ_1 and the $\mathcal{H}_2$ norms of the closed loop for the multi-input multi-output (MIMO) case which is much richer in its complexity and its applicability than the SISO case, discussed in previous chapters. Consider for example Figure 5.1 where the part of the regulated output given by z_2 is used to reflect the performance with respect to a unit pulse input and it is also required that the maximum magnitude of z_1 due to a worst unity magnitude bounded input stays below a prespecified level. This objective can be captured by the following problem:

$$\min_{K \text{ stabilizing}} \{||\Phi_{z_2 w}||_2 : ||\Phi_{z_1 w}||_1 \leq \gamma\}, \tag{7.1}$$

where γ is the level over which the infinity norm of z_2 is not allowed to cross for the worst bounded disturbance. Or, it may be that the disturbance w is such that a part of it, w_1 is a white noise while another part, w_2 is unity magnitude bounded. A relevant objective is the minimization of the effect of these disturbances on the regulated output. The problem,

$$\min_{K \text{ stabilizing}} \{||\Phi_{z w_1}||_2 : ||\Phi_{z w_2}||_1 \leq \gamma\}, \tag{7.2}$$

where γ is the level over which the infinity norm of z is not allowed to cross for the worst unity magnitude bounded disturbance is then a problem of interest.

Both problems mentioned previously fall under a general framework of a problem which we call the *mixed problem.* This problem is addressed and solved in this chapter along with a related problem which minimizes a combination of the various input-output maps of the closed loop which we call the *combination problem.* The latter is of interest by itself and in relation to the mixed problem. The treatment in the chapter is restricted to the square case also known as the 1-block case (where the number of regulated variables is equal to the number of control inputs and the number of measured outputs is equal to the number of exogenous inputs).

The MIMO problem in the mixed $\mathcal{H}_2$ and ℓ_1 setting poses many questions which are not addressed in the SISO setting discussed in previous chapters. It is shown that MIMO problems need to be handled differently in a significant way. The optimal solutions for the 1-block are not in general finite impulse responses nor are they unique (as will be shown) unlike the SISO case. However, it is established that the solution can be obtained via finite dimensional quadratic optimization and linear programming. We show that it is possible to obtain an *a priori* bound on the dimension of the suboptimal problems even for the MIMO version of the problem addressed in [14] (no *a priori* bound is given in [14]).

The chapter is organized as follows. In section 7.1 we give preliminaries. In section 7.2 we show that the combination problem can be solved exactly via a finite dimensional quadratic optimization and linear programming for the square case. In section 7.3 we study the mixed problem and its relation to the combination problem. In section 7.4 we give an example to illustrate the theory developed. In section 7.5 conclusions are given. Section 7.6 is the appendix which contains the proofs of some of the facts stated in earlier sections.

7.1 Preliminaries

In this section we state theorems, assumptions and give notation which will be relevant to the rest of the chapter. A good reference for this section is [3]. We denote by n_u, n_w, n_z and n_y the number of control inputs, exogenous inputs, regulated outputs and measured outputs respectively of the plant G. We represent by Θ, the set of closed loop maps of the plant G which are achievable through stabilizing controllers. $H \in \ell_1^{n_z \times n_w}$, $U \in \ell_1^{n_z \times n_u}$ and $V \in \ell_1^{n_y \times n_w}$ characterize the Youla parametrization of the plant. The following theorem follows from the Youla parametrization.

Theorem 7.1.1.

$$\Theta = \{\Phi \in \ell_1^{n_z \times n_w} : \text{there exists a } Q \in \ell_1^{n_u \times n_y} \text{ with } \hat{\Phi} = \hat{H} - \hat{U}\hat{Q}\hat{V}\}.$$

If Φ is in Θ we say that Φ is an *achievable* closed loop map. We assume throughout the chapter that *$\hat{U}$ has normal rank n_u and $\hat{V}$ has normal rank n_y*. There is no loss of generality in making this assumption [3]. Let the Smith-McMillan decomposition of $\hat{U}$ and $\hat{V}$ be given by

$$\begin{aligned}\hat{U} &= \hat{L}_U \hat{M}_U \hat{R}_U,\\ \hat{V} &= \hat{L}_V \hat{M}_V \hat{R}_V,\end{aligned}$$

respectively where $L_U \in \ell_1^{n_z \times n_z}$, $R_U \in \ell_1^{n_u \times n_u}$, $L_V \in \ell_1^{n_y \times n_y}$ and $R_V \in \ell_1^{n_w \times n_w}$ are unimodular matrices. Let Λ_{UV} denote the set of zeros of $\hat{U}$ and $\hat{V}$ in $\mathcal{D}$ (i.e. the zeros of $\Pi_{i=1}^{n_u} \hat{\epsilon}_i \Pi_{j=1}^{n_y} \hat{\epsilon}_j'$ which lie inside the closed unit disc). For $\lambda_0 \in \Lambda_{UV}$ define,

$$\begin{aligned}\sigma_{U_i}(\lambda_0) &:= \text{multiplicity of } \lambda_0 \text{ as a root of } \hat{\epsilon}_i(\lambda) \text{ for } i = 1, \ldots, n_u,\\ \sigma_{V_j}(\lambda_0) &:= \text{multiplicity of } \lambda_0 \text{ as a root of } \hat{\epsilon}_j'(\lambda) \text{ for } j = 1, \ldots, n_y.\end{aligned}$$

As $\hat{L}_U$, $\hat{R}_V$ are unimodular we can define the following polynomial row and column vectors:

$$\hat{\alpha}_i(\lambda) = (\hat{L}_U^{-1})_i(\lambda) \text{ for } i = 1, \ldots, n_z,$$
$$\hat{\beta}_j(\lambda) = (\hat{R}_V^{-1})^j(\lambda) \text{ for } j = 1, \ldots, n_w,$$

where $(M)_i$ denotes the ith row of the matrix M and $(M)^j$ denotes the jth column of a matrix M. Note that $\alpha_i \in \ell_1^{1 \times n_z}$ and $\beta_j \in \ell_1^{n_w \times 1}$. We assume that $\hat{U}$ *and* $\hat{V}$ *have no zeros which lie on the unit circle.* This is a standard assumption in the optimal model-matching approach we employ and is crucial for the 1-block development. For a detailed discussion about the case where this assumption fails see [3]. We now present the main interpolation theorem for a closed loop map to be achievable. We denote the kth derivative of (.) with respect to λ by $(.)^{(k)}$ whereas the kth power of (.) is denoted by $(.)^k$.

Theorem 7.1.2. *[3] $\Phi \in \ell_1^{n_z \times n_w}$ is in Θ if and only if the following conditions hold for all $\lambda_0 \in \Lambda_{UV}$*

$$i)(\alpha_i \hat{\Phi} \hat{\beta}_j)^{(k)}(\lambda_0) = (\alpha_i \hat{H} \hat{\beta}_j)^{(k)}(\lambda_0) \ for \begin{cases} i = 1, \ldots, n_u \\ j = 1, \ldots, n_y \\ k = 0, \ldots, \sigma_{U_i}(\lambda_0) + \sigma_{V_j}(\lambda_0) - 1 \end{cases}$$

$$ii) \begin{cases} (\hat{\alpha}_i \hat{\Phi})(\lambda) = (\hat{\alpha}_i \hat{H})(\lambda) \ for \ i = n_u + 1, \ldots, n_z \\ (\hat{\Phi} \hat{\beta}_j)(\lambda) = (\hat{H} \hat{\beta}_j)(\lambda) \ for \ j = n_y + 1, \ldots, n_w \end{cases}.$$

The first set of conditions constitutes the *zero interpolation* conditions whereas the second set consists of the *rank interpolation conditions.* The plant G is called *square,* or equivalently, we have a 1-block problem, if the rank interpolation conditions are absent (i.e., when $n_u = n_z$ and $n_y = n_w$). Otherwise, the plant is *non-square plant,* or equivalently, we have a 4-block problem. Define $F^{ijk\lambda_0} \in \ell_\infty^{n_z \times n_w}$ by

$$F_{pq}^{ijk\lambda_0}(s) := \sum_{l=0}^{\infty} \sum_{t=0}^{\infty} \alpha_{ip}(t-l)\beta_{jq}(l-s) \, (\lambda^t)^{(k)}\Big|_{\lambda=\lambda_0} . \tag{7.3}$$

It can be easily verified that for any $\Phi \in \ell_1^{n_z \times n_w}$,

$$(\hat{\alpha}_i \hat{\Phi} \hat{\beta}_j)^{(k)}(\lambda_0) = \langle \Phi, F^{ijk\lambda_0} \rangle . \tag{7.4}$$

This is shown in the Appendix. $F^{ijk\lambda_0}$ characterize the zero interpolation conditions. We state the following lemma which will be of use in the next section.

Lemma 7.1.1. *For all $\lambda_0 \in \Lambda_{UV}$:*

$$F^{ijk\lambda_0} \in \ell_1^{n_z \times n_w} \ (and\ hence \in \ell_2^{n_z \times n_w}) \ for \begin{cases} i = 1, \ldots, n_u \\ j = 1, \ldots, n_y \\ k = 0, \ldots, \sigma_{U_i}(\lambda_0) + \sigma_{V_j}(\lambda_0) \\ \qquad -1. \end{cases}$$

Furthermore, given any $\epsilon > 0$ we can always choose a $T_0 \geq T$ such that

$$|F_{pq}^{ijk\lambda_0}(s)| < \epsilon \ for\ all\ s > T_0.$$

Proof. See Appendix.

This means that the zero interpolation conditions can be characterized via elements in $\ell_1^{n_z \times n_w}$. Similarly, the rank interpolation conditions can be characterized via elements $G_{\alpha_i qt}$ and $G_{\beta_j pt}$ in $\ell_1^{n_z \times n_w}$ (see Appendix) as the following theorem states. In the following theorem $G_{\alpha_i qt}$ and $G_{\beta_j pt}$ (elements in $\ell_1^{n_z \times n_w}$) are defined in the Appendix.

Theorem 7.1.3. *[3] Let*

$$RF^{ijk\lambda_0} := Real(F^{ijk\lambda_0}) \ and \ IF^{ijk\lambda_0} = Imaginary(F^{ijk\lambda_0}).$$

Suppose $\Lambda_{UV} \subset int(\mathcal{D})$ then $\Phi \in \ell_1^{n_z \times n_w}$ is in Θ if and only if the following conditions hold:

$$\begin{array}{l} <\Phi, RF^{ijk\lambda_0}> = <H, RF^{ijk\lambda_0}> \\ <\Phi, IF^{ijk\lambda_0}> = <H, IF^{ijk\lambda_0}> \end{array} \ for \begin{cases} \lambda_0 \in \Lambda_{UV} \\ i = 1, \ldots, n_u \\ j = 1, \ldots, n_y \\ k = 0, \ldots, \sigma_{U_i}(\lambda_0) + \sigma_{V_j}(\lambda_0) - 1 \end{cases}$$

and

$$\begin{array}{l} <\Phi, G_{\alpha_i qt}> = <H, G_{\alpha_i qt}> \\ <\Phi, G_{\beta_j pt}> = <H, G_{\beta_j pt}> \end{array} \ for \begin{cases} i = n_u + 1, \ldots, n_z \\ j = n_y + 1, \ldots, n_w \\ q = 1, \ldots, n_w \\ p = 1, \ldots, n_z \\ t = 0, 1, 2, \ldots \end{cases}$$

Furthermore, $F^{ijk\lambda_0}, G_{\alpha_i qt}$ and $G_{\beta_j pt}$ are matrix sequences in $\ell_1^{n_z \times n_w}$.

Proof. Follows easily from Theorem 7.1.2, equation (7.4) and the fact that H and R are real matrix sequences. The fact about sequences in $\ell_1^{n_z \times n_w}$ is shown in the Appendix. □

We assume without loss of generality that $F^{ijk\lambda_0}$ is a real sequence. Further, we define

$$b^{ijk\lambda_0} := <H, F^{ijk\lambda_0}> \ \text{and} \ c_z := \sum_{\lambda_0 \in \Lambda_{UV}} \sum_{i=1}^{n_u} \sum_{j=1}^{n_y} \sigma_{U_i}(\lambda_0) + \sigma_{V_j}(\lambda_0).$$

c_z is the total number of zero interpolation conditions. The following problem

$$\nu_{0,1} = \inf_{\Phi \ \text{Achievable}} \{||\Phi||_1\}, \tag{7.5}$$

is the standard multiple input multiple output $\ell_1^{n_z \times n_w}$ problem. In [5] it is shown that this problem for a square plant has a solution, possibly nonunique but the solution is a finite impulse response matrix sequence. Let

$$\mu_{0,2} := \inf_{\Phi \ \text{Achievable}} \{||\Phi||_2^2\}, \tag{7.6}$$

which is the standard $\mathcal{H}_2$ problem. The solution to this problem is unique and is an infinite impulse response sequence. We now collect all the assumptions made (which will be assumed throughout this chapter) for easy reference.

Assumption 2 *$\hat{U}$ has normal rank n_u and $\hat{V}$ has normal rank n_y.*

Assumption 3 *$\hat{U}$ and $\hat{V}$ have no zeros which lie on the unit circle, that is $\Lambda_{UV} \subset int(\mathcal{D})$.*

Assumption 4 *$F^{ijk\lambda_0}$ is a real sequence.*

7.2 The Combination Problem

In this section we state and solve the combination problem. We first make the problem statement precise. Next we show the existence of an optimal solution. We then solve the problem for the square case. Finally, we study the nonsquare case.

Let $N_w := \{1, \ldots, n_w\}$ and let $N_z := \{1, \ldots, n_z\}$. Let M, N and MN be subsets of $N_z \times N_w$ such that the intersection between any two of these sets is empty and their union is $N_z \times N_w$. Let $\bar{c}_{pq}$ and c_{pq} be given positive constants for $(p,q) \in MN \cup M$ and for $(p,q) \in MN \cup N$ respectively. The problem of interest is the following: *Given a plant G solve the following optimization problem:*

$$\nu = \inf_{\Phi \text{ Achievable}} \sum_{(p,q)\in MN\cup M} \bar{c}_{pq}\|\Phi_{pq}\|_2^{\,2} + \sum_{(p,q)\in MN\cup N} c_{pq}\|\Phi_{pq}\|_1. \quad (7.7)$$

Note that for all $(p,q) \in M$ only the $\mathcal{H}_2$ norm of Φ_{pq} appears in the objective, for all $(p,q) \in N$ only the ℓ_1 norm of Φ_{pq} appears in the objective and for all $(p,q) \in MN$ a combination of the $\mathcal{H}_2$ and the ℓ_1 norm of Φ_{pq} appears in the objective. For notational convenience we define $f : \ell_2^{n_z \times n_w} \to R$ by

$$f(\Phi) := \sum_{(p,q)\in MN\cup M} \bar{c}_{pq}\|\Phi_{pq}\|_2^{\,2} + \sum_{(p,q)\in MN\cup N} c_{pq}\|\Phi_{pq}\|_1,$$

which is the objective functional being minimized. As it can be seen the objective functional of the combination problem constitutes a weighted sum of the square of the $\mathcal{H}_2$ norm and the ℓ_1 norm of individual elements Φ_{pq} of the closed loop map Φ. Note that with this type of functional the overall $\mathcal{H}_2$ norm of the closed loop as well as ℓ_1 norms of individual rows can be incorporated as special cases.

For technical reasons explained in the sequel we define the space

$$\mathcal{A} := \{\Phi \in \ell_2^{n_z \times n_w} : \Phi_{pq} \in \ell_1 \text{ for all } (p,q) \in MN \cup N\}.$$

The following set is an extension of Θ

$\Theta_e := \{\Phi \in \mathcal{A} :\ \Phi$ satisfies the zero and the rank interpolation conditions$\}$.

Note that Θ is the set

$\{\Phi \in \ell_1^{n_z \times n_w} :\ \Phi$ satisfies the zero and the rank interpolation conditions$\}$.

Also, note that when M is empty then $\Theta = \Theta_e$. Finally, we define the following optimization problem

$$\nu_e := \inf_{\Phi \in \Theta_e} f(\Phi). \quad (7.8)$$

Now, we show that a solution to (7.8) always exists.

Lemma 7.2.1. *There exists $\Phi^0 \in \Theta_e$ such that*

$$\nu_e = \sum_{(p,q)\in MN\cup M} \bar{c}_{pq}\|\Phi^0_{pq}\|_2^{\,2} + \sum_{(p,q)\in MN\cup N} c_{pq}\|\Phi^0_{pq}\|_1.$$

Therefore, the infimum in (7.8) is a minimum.

Proof. See Appendix.

7.2.1 Square Case

Here, we solve the combination problem for the square case. Throughout this subsection the following assumption holds:

Assumption 5 $n_u = n_z$ *and* $n_y = n_w$.

In the sequel $y \in R^{c_z}$ is indexed by $ijk\lambda_0$ where i, j, k, λ_0 vary as in the zero interpolation conditions. The following lemma gives the dual problem for the square case.

Lemma 7.2.2.

$$\nu_e = \max\{\psi(y),\ y \in R^{c_z}\}, \tag{7.9}$$

where $\psi(y) = \inf_{\Phi\in\mathcal{A}} L(\Phi)$ *and*

$$L(\Phi) := \sum_{(p,q)\in(MN)\cup M} \bar{c}_{pq}\|\Phi_{pq}\|_2^{\,2} + \sum_{(p,q)\in(MN)\cup N} c_{pq}\|\Phi_{pq}\|_1 + \sum_{i,j,k,\lambda_0} y_{ijk\lambda_0}(b^{ijk\lambda_0} - \langle F^{ijk\lambda_0}, \Phi\rangle).$$

Proof. We will apply Theorem 3.3.2 (Kuhn-Tucker-Lagrange duality theorem) to get the result. Let X, Ω, Y, Z in Theorem 3.3.2 correspond to $\mathcal{A}, \mathcal{A}, R^{c_z}, R$ respectively. Let $\gamma = \nu_e + 1$ and let $g : \mathcal{A} \to R$ be given by $g(\Phi) := f(\Phi) - \gamma$. Let $\underline{H} : \mathcal{A} \to R^{c_z}$ be given by

$$\underline{H}_{ijk\lambda_0}(\Phi) := b^{ijk\lambda_0} - \langle F^{ijk\lambda_0}, \Phi\rangle.$$

We index the equality constraints of $\underline{H}$ by $ijk\lambda_0$ where i, j, k, λ_0 vary as in the zero interpolation conditions. In [3] it is shown that the map $\underline{H}$ is onto R^{c_z} (it is shown that the zero interpolation conditions are independent). This means that $0 \in int(Range(\underline{H}))$. From Lemma 7.2.1 we know that there exists a $\Phi^1 \in \mathcal{A}$ such that $\underline{H}(\Phi^1) = 0$ (that is Φ^1 satisfies the zero interpolation conditions) and $f(\Phi^1) = \nu_e$ which implies that $g(\Phi^1) = -1 < 0$. Thus all the conditions of Theorem 3.3.2 (Kuhn-Tucker-Lagrange duality theorem) are saitisfied. The lemma follows by applying Theorem 3.3.2 to (7.8). □

For notational convenience we define the functionals $Z_{pq} \in \ell_1$ by

$$Z_{pq}(t) := \sum_{i,j,k,\lambda_0} y_{ijk\lambda_0} F^{ijk\lambda_0}_{pq}(t).$$

In what follows we show that the dual problem is in fact a finite dimensional convex programming problem. A bound on its dimension is also furnished.

Theorem 7.2.1. *It is true that $\nu = \nu_e$. ν_e can be obtained by solving the following problem:*

$$\max\{\sum_{(p,q)\in M}\sum_{t=0}^{\infty} -\bar{c}_{pq}\Phi_{pq}(t)^2 + \sum_{(p,q)\in MN}\sum_{t=0}^{\infty} -\bar{c}_{pq}\Phi_{pq}(t)^2 + \sum_{i,j,k,\lambda_0} y_{ijk\lambda_0} b^{ijk\lambda_0}\}$$

subject to

$$\left.\begin{array}{ll} y \in R^{c_s}, \Phi_{pq} \in \ell_1 \text{ for all } (p,q) \in MN \cup M, & \\ -c_{pq} \le Z_{pq}(t) \le c_{pq} & \text{if } (p,q) \in N \\ 2\bar{c}_{pq}\Phi_{pq}(t) = Z_{pq}(t) - c_{pq} & \text{if } (p,q) \in MN \text{ and } Z_{pq}(t) > c_{pq}, \\ \quad = Z_{pq}(t) + c_{pq} & \text{if } (p,q) \in MN \text{ and } Z_{pq}(t) < -c_{pq}, \\ \quad = 0 & \text{if } (p,q) \in MN \text{ and } |Z_{pq}(t)| \le c_{pq}, \\ \quad = Z_{pq}(t) & \text{if } (p,q) \in M, \\ \text{for all } t = 0,1,2,\ldots. & \end{array}\right\} (I)$$

Furthermore, it holds that the infimum in (7.7) is a minimum, and, Φ^0 is a solution of (7.8) if and only if it is a solution of (7.7). In addition, Φ^0_{pq} is unique for all $(p,q) \in (MN) \cup M$.

Proof. In Lemma 7.2.1 we showed that an optimal solution Φ^0 always exists for problem (7.8). From Theorem 3.3.2 (Kuhn-Tucker-Lagrange duality theorem) we know that if $y \in R^{c_s}$ is optimal for the dual problem then Φ^0 minimizes $L(\Phi)$ where

$$L(\Phi) = \sum_{(p,q)\in(MN)\cup M} \bar{c}_{pq}||\Phi_{pq}||_2{}^2 + \sum_{(p,q)\in(MN)\cup N} c_{pq}||\Phi_{pq}||_1$$
$$+ \sum_{i,j,k,\lambda_0} y_{ijk\lambda_0}(b^{ijk\lambda_0} - < F^{ijk\lambda_0}, \Phi >).$$

Thus, $\Phi^0(t)$ minimizes

$$\sum_{(p,q)\in M} (\bar{c}_{pq}\Phi_{pq}(t)^2 - Z_{pq}(t)\Phi_{pq}(t)) + \sum_{(p,q)\in MN} (\bar{c}_{pq}\Phi_{pq}(t)^2 + c_{pq}|\Phi_{pq}(t)|$$
$$-Z_{pq}(t)\Phi_{pq}(t)) + \sum_{(p,q)\in N} (c_{pq}|\Phi_{pq}(t)| - Z_{pq}(t)\Phi_{pq}(t)) + \sum_{i,j,k,\lambda_0} y_{ijk\lambda_0} b^{ijk\lambda_0}.$$

Therefore, if $(p,q) \in M$ then $\Phi^0_{pq}(t)$ minimizes

$$\bar{c}_{pq}\Phi_{pq}(t)^2 - Z_{pq}(t)\Phi_{pq}(t),$$

which is strictly convex in $\Phi_{pq}(t)$ and therefore $\Phi^0_{pq}(t)$ is unique. Differentiating the above function with respect to $\Phi_{pq}(t)$ and equating the result to zero we conclude that if $(p,q) \in M$ then

$$2\bar{c}_{pq}\Phi^0_{pq}(t) = Z_{pq}(t).$$

As $Z_{pq} \in \ell_1$ we have that for all $(p,q) \in M, \Phi^0_{pq} \in \ell_1$.

If $(p,q) \in MN$ then $\Phi^0_{pq}(t)$ minimizes

$$\bar{c}_{pq}\Phi_{pq}(t)^2 + c_{pq}|\Phi_{pq}(t)| - Z_{pq}(t)\Phi_{pq}(t),$$

which is strictly convex and therefore the minimizer is unique. This also implies that if $(p,q) \in MN$ then $\Phi^0_{pq}(t)$ satisfies conditions stipulated in (I).

Suppose, $(p,q) \in N$ then $\Phi^0_{pq}(t)$ minimizes

$$\overline{f}(\Phi_{pq}(t)) := \{c_{pq}|\Phi_{pq}(t)| - Z_{pq}(t)\Phi_{pq}(t)\}.$$

Note that if $Z_{pq}(t)\Phi_{pq}(t) < 0$ then $\overline{f}(\Phi_{pq}(t)) \geq 0$. But, $\overline{f}(0) = 0$. Therefore, the optimal minimizes

$$\Phi_{pq}(t)(c_{pq}sgn(Z_{pq}(t)) - Z_{pq}(t)),$$

over all $\Phi_{pq}(t) \in R$ such that $Z_{pq}(t)\Phi_{pq}(t) \geq 0$. Now, if $|Z_{pq}(t)| > c_{pq}$ then given any $K > 0$ we can choose $\Phi_{pq}(t) \in R$ that satisfies $Z_{pq}(t)\Phi_{pq}(t) > 0$ and $\overline{f}(\Phi_{pq}(t)) < -K$ and thus the infimum value would be $-\infty$. Therefore, we can restrict $Z_{pq}(t)$ in the maximization of the dual to satisfy $|Z_{pq}(t)| \leq c_{pq}$. If, $|Z_{pq}(t)| < c_{pq}$ then $\overline{f}(\Phi_{pq}(t)) \geq 0$ for any $\Phi_{pq}(t) \in R$ that satisfies $Z_{pq}(t)\Phi_{pq}(t) \geq 0$ and is equal to zero *only* for $\Phi_{pq}(t) = 0$. Therefore, we conclude that if $(p,q) \in N$ then we can restrict $Z_{pq}(t)$ in the maximization of the dual to satisfy $|Z_{pq}(t)| \leq c_{pq}$ and that the optimal $\Phi^0_{pq}(t)$ minimizes $\overline{f}(\Phi_{pq}(t))$ with a minimum value of zero. It also follows that if $|Z_{pq}(t)| < c_{pq}$ then $\Phi_{pq}(t) \in R$ that minimizes $\overline{f}(\Phi_{pq}(t))$ is equal to zero. The expression for ν_e follows by substituting the value of $\Phi^0_{pq}(t)$ obtained in the above discussion for various indices in the functional $L(\Phi)$.

Note that $\Theta = \Theta_e \cap \ell_1^{n_z \times n_w}$. But in the previous steps we have shown by construction that the optimal solution to (7.8), Φ^0, is such that $\Phi^0_{pq} \in \ell_1$ for all $(p,q) \in M$. This means that $\Phi^0 \in \Theta$. From the above discussion the theorem follows easily. □

Note that the above theorem demonstrates that the problem at hand is finite dimensional. Indeed, at an optimal point y^0, Φ^0 the constraint $|Z^0_{pq}(t)| \leq c_{pq}$ will be satisfied for sufficiently large t since $Z^0_{pq} \in \ell_1$. Thus, $\Phi^0_{pq}(t) = 0$ for $(p,q) \in MN \cup N$ and large t i.e., Φ^0_{pq} is FIR for $(p,q) \in MN \cup N$. The following lemmas provide a way to compute *a priori* bounds on the dimension of the problem.

Lemma 7.2.3. *Let Φ^0 be a solution to the primal problem (7.7) and let y^0, Z^0_{pq} be solutions to the dual. Then the following is true:*

$$\begin{aligned}
-c_{pq} \leq Z^0_{pq}(t) &\leq c_{pq} && if\ (p,q) \in N,\\
\Phi^0_{pq}(t) &= 0 && if\ (p,q) \in N\ and\ |Z^0_{pq}(t)| < c_{pq},\\
2\overline{c}_{pq}\Phi^0_{pq}(t) &= Z^0_{pq}(t) - c_{pq} && if\ (p,q) \in MN\ and\ Z^0_{pq}(t) > c_{pq},\\
&= Z^0_{pq}(t) + c_{pq} && if\ (p,q) \in MN\ and\ Z^0_{pq}(t) < -c_{pq},\\
&= 0 && if\ (p,q) \in MN\ and\ |Z^0_{pq}(t)| \leq c_{pq},\\
&= Z^0_{pq}(t) && if\ (p,q) \in M.
\end{aligned}$$

Φ^0_{pq} is unique for all (p,q) in $(MN) \cup M$. Also, there exists an a priori bound α such that $||Z^0_{pq}||_\infty \leq \alpha$ for all $(p,q) \in N_z \times N_w$.

Proof. The first part of the lemma follows from the arguments used in Theorem 7.2.1. We now determine an *a priori* bound. For all $(p,q) \in MN$ the following is true:

$$\begin{aligned}|Z^0_{pq}(t)| &\le c_{pq} + 2\bar{c}_{pq}|\Phi^0_{pq}(t)| \le c_{pq} + 2\bar{c}_{pq}||\Phi^0_{pq}||_1 \\ &\le c_{pq} + \tfrac{2\bar{c}_{pq}}{c_{pq}} f(H)\end{aligned}$$

where the last inequality follows since H is a feasible solution and hence $c_{pq}||\Phi^0_{pq}||_1 \le f(\Phi^0) \le f(H)$. For all $(p,q) \in N$ the following is true:

$$|Z^0_{pq}(t)| \le c_{pq}.$$

For all $(p,q) \in M$

$$\begin{aligned}|Z^0_{pq}(t)| &\le 2\bar{c}_{pq}|\Phi^0_{pq}(t)| \le 2\bar{c}_{pq}||\Phi^0_{pq}||_2 \\ &\le 2\bar{c}_{pq}\sqrt{\tfrac{f(\Phi^0)}{\bar{c}_{pq}}} \le 2\bar{c}_{pq}\sqrt{\tfrac{f(H)}{\bar{c}_{pq}}}.\end{aligned}$$

Denote by d_{pq} the upper bounds determined above, all of which can be determined *a priori*. Let

$$\alpha := \max_{(p,q)\in N_z \times N_w} d_{pq}.$$

Thus α is an *a priori* upperbound on $||Z^0_{pq}||_\infty$ for all $(p,q) \in N_z \times N_w$. This proves the lemma. □

Lemma 7.2.4. *Let*

$$c_{min} := \min_{(p,q)\in M N \cup N} c_{pq}.$$

If $y \in R^{c_z}$ *is such that for all* $(p,q) \in N_z \times N_w$

$$||Z_{pq}||_\infty \le \alpha,$$

where

$$Z_{pq}(t) := \sum_{ijk\lambda_0} y_{ijk\lambda_0} F^{ijk\lambda_0}_{pq}(t),$$

then there exists an a priori computable positive integer L^* *such that*

$$|Z_{pq}(t)| < c_{min} \textit{ for all } t \ge L^*.$$

Proof. For notational convenience we index $F^{ijk\lambda_0}_{pq}$ and $b^{ijk\lambda_0}$ where $ijk\lambda_0$ vary as in the zero interpolation conditions by F^n_{pq} and b^n respectively where $n = 1, \ldots, c_z$. The vector in R^{c_z} whose n^{th} element is given by b^n is denoted by b. We interpret F^n_{pq} as a $\infty \times 1$ column vector equal to

$$(F^n_{pq}(0), F^n_{pq}(1), \ldots)'.$$

With this notation

$$Z_{pq} = (F^1_{pq}, F^2_{pq}, \ldots, F^{c_z}_{pq})y,$$

where Z_{pq} is viewed as a infinite column vector with the tth element equal to $Z_{pq}(t)$. Therefore the condition

$$||Z_{pq}||_\infty \leq \alpha \text{ for all } (p,q) \in N_z \times N_w,$$

is equivalent to the condition

$$||A'y||_\infty \leq \alpha,$$

where

$$A' = \begin{pmatrix} F_{11}^1 & F_{11}^2 & \dots & F_{11}^{c_z} \\ F_{12}^1 & F_{12}^2 & \dots & F_{12}^{c_z} \\ \vdots & \vdots & \vdots & \vdots \\ F_{1n_w}^1 & F_{1n_w}^2 & \dots & F_{1n_w}^{c_z} \\ F_{21}^1 & F_{21}^2 & \dots & F_{21}^{c_z} \\ \vdots & \vdots & \vdots & \vdots \\ F_{n_z n_w}^1 & F_{n_z n_w}^2 & \dots & F_{n_z n_w}^{c_z} \end{pmatrix}.$$

The matrix $A := (A')'$ is the matrix which has c_z rows each for one zero interpolation condition. If $\Phi \in \ell_1^{n_z \times n_w}$ is stringed out into a vector as below:

$$\Phi = \begin{pmatrix} \Phi_{11} \\ \Phi_{12} \\ \vdots \\ \Phi_{1n_w} \\ \Phi_{21} \\ \vdots \\ \Phi_{n_z n_w} \end{pmatrix},$$

then $A\Phi = b$ gives the zero interpolation conditions. It is known that the zero interpolation conditions are independent and therefore A has full row rank. Equivalently, A' has full column rank. Choose $L \geq c_z$ such that $D \in R^{c_z \times c_z}$ with rows from the first L rows of A' is invertible. Consider D as a map from $(R^{c_z}, |.|_1)$ to $(R^{c_z}, |.|_\infty)$. Now, as $y = D^{-1}Dy$ we have

$$|y|_1 = |D^{-1}Dy|_1 \leq |D^{-1}|_{\infty,1}|Dy|_\infty \leq |D^{-1}|_{\infty,1}\alpha, \tag{7.10}$$

where $|D^{-1}|_{\infty,1}$ is the induced norm of D^{-1}. The last inequality follows because $||A'y||_\infty \leq \alpha$ which implies $|Dy|_\infty \leq \alpha$.

From the note after Lemma 7.1.1 we know that for any n and for any $(p,q) \in N_z \times N_w$ there exists an integer T_{pqn} such that $t > T_{pqn}$ implies that

$$|F_{pq}^n(t)| < \frac{c_{min}}{|D^{-1}|_{\infty,1}\alpha}.$$

Let

$$L^* = \max_{pqn} T_{pqn}.$$

Note that L^* is determined *a priori*. Let $t > L^*$ then

$$|Z_{pq}(t)| = |\sum_{n=1}^{c_s} F_{pq}^n(t) y_n| \leq \sum_{n=1}^{c_s} |y_n| |F_{pq}^n(t)|$$
$$\leq \frac{c_{min}}{|D^{-1}|_{\infty,1}\alpha} \sum_{n=1}^{c_s} |y_n| \leq c_{min}.$$

The last inequality follows from equation (7.10). This proves the lemma. □

We now state the main result of this subsection.

Theorem 7.2.2. *It is true that ν equals*

$$\max\{ \sum_{(p,q)\in M} \sum_{t=0}^{\infty} -\frac{1}{4\bar{c}_{pq}} Z_{pq}(t)^2 + \sum_{(p,q)\in MN} \sum_{t=0}^{L^*} -\bar{c}_{pq}\Phi_{pq}(t)^2 + \sum_{i,j,k,\lambda_0} y_{ijk\lambda_0} b^{ijk\lambda_0}\}$$

subject to

$$\left.\begin{array}{ll} y \in R^{c_s}, \Phi_{pq} \in R^{L^*} \text{ for all } (p,q) \in MN, & \\ -c_{pq} \leq Z_{pq}(t) \leq c_{pq} & \text{if } (p,q) \in N \\ -c_{pq} \leq 2\bar{c}_{pq}\Phi_{pq}(t) - Z_{pq}(t) \leq c_{pq} & \text{if } (p,q) \in MN \end{array}\right\} (III)$$

for all $t = 0, 1, 2, \ldots, L^*$ where L^* is determined a priori as given in Lemma 7.2.4.

Furthermore, the optimal of the primal (7.7) Φ_{pq}^0 is unique for all $(p,q) \in (MN) \cup M$.

Proof. An easy conclusion of Lemma 7.2.3 and Lemma 7.2.4 is that ν is equal to

$$\max\{ \sum_{(p,q)\in M} \sum_{t=0}^{\infty} -\frac{1}{4\bar{c}_{pq}} Z_{pq}(t)^2 + \sum_{(p,q)\in MN} \sum_{t=0}^{L^*} -\bar{c}_{pq}\Phi_{pq}(t)^2 + \sum_{i,j,k,\lambda_0} y_{ijk\lambda_0} b^{ijk\lambda_0}\}$$

subject to

$$\left.\begin{array}{rll} & y \in R^{c_s}, \Phi_{pq} \in R^{L^*} \text{ for all } (p,q) \in MN, & \\ -c_{pq} \leq & Z_{pq}(t) \leq c_{pq} & \text{if } (p,q) \in N \\ 2\bar{c}_{pq}\Phi_{pq}(t) = & Z_{pq}(t) - c_{pq} & \text{if } (p,q) \in MN \text{ and } Z_{pq}(t) > c_{pq}, \\ = & Z_{pq}(t) + c_{pq} & \text{if } (p,q) \in MN \text{ and } Z_{pq}(t) < -c_{pq}, \\ = & 0 & \text{if } (p,q) \in MN \text{ and } |Z_{pq}(t)| \leq c_{pq}, \end{array}\right\} (II)$$
$$\textit{for all } t = 0, 1, 2, \ldots, L^*.$$

L^* is determined *a priori* as indicated in Lemma 7.2.4. Indeed, from Lemma 7.2.3 we know that we can restrict the maximization in Theorem 7.2.1 over the set $y \in R^{c_s}$ such that $||Z_{pq}||_\infty \leq \alpha$. Now we conclude from Lemma 7.2.4 that there exists an integer L^* such that $t > L^*$ implies that $|Z_{pq}(t)| < c_{min} \leq c_{pq}$. One of the conditions stipulated in (I) of Theorem 7.2.1 is that

$$\Phi_{pq}(t) = 0 \text{ if } (p,q) \in MN \text{ and } |Z_{pq}(t)| < c_{pq}.$$

Note that the condition $2\Phi_{pq}(t) = Z_{pq}(t)$ for $(p,q) \in M$ has been incorporated into the objective functional of this theorem. The fact that Φ_{pq}^0 is unique for all $(p,q) \in (MN) \cup M$ is due to Theorem 7.2.1.

To bring the problem into the form stated in the theorem, denote the right hand side of the equation in the statement of the theorem by $\bar{\nu}$. Suppose,

$y \in R^{c_s}$, $Z_{pq}(t)$ (determined by y), and $\Phi_{pq}(t)$ satisfy condition (II) above for all the appropriate indices. If $(p,q) \in MN$ and $Z_{pq}(t) > c_{pq}$ then

$$-c_{pq} = 2\bar{c}_{pq}\Phi_{pq}(t) - Z_{pq}(t) \leq c_{pq},$$

because $c_{pq} \geq 0$. Similarly it can be checked that all the other conditions of (III) are satisfied. This implies that $\bar{\nu} \geq \nu$. Suppose, $y \in R^{c_s}$, $Z_{pq}(t)$ (determined by y) and $\Phi_{pq}(t)$ satisfy condition (III) of Theorem 7.2.2. Let $\overline{\Phi}_{pq}(t)$ be defined as follows:

$2\bar{c}_{pq}\overline{\Phi}_{pq}(t) = Z_{pq}(t) - c_{pq}$ if $(p,q) \in MN$ and $Z_{pq}(t) > c_{pq}$,
$2\bar{c}_{pq}\overline{\Phi}_{pq}(t) = Z_{pq}(t) + c_{pq}$ if $(p,q) \in MN$ and $Z_{pq}(t) < -c_{pq}$,
$\overline{\Phi}_{pq}(t) = 0$ if $(p,q) \in MN$ and $|Z_{pq}(t)| \leq c_{pq}$,
for all $0 \leq t \leq L^*$ (i.e $\overline{\Phi}_{pq}(t)$ satisfies constraints (II)).
Suppose, $(p,q) \in MN$ and $Z_{pq}(t) > c_{pq}$ then

$$0 \leq 2\bar{c}_{pq}\overline{\Phi}_{pq}(t) = Z_{pq}(t) - c_{pq} \leq 2\bar{c}_{pq}\Phi_{pq}(t).$$

Therefore,

$$-\overline{\Phi}^2_{pq}(t) \geq -\Phi^2_{pq}(t).$$

Similarly, the above condition follows for other indices. Thus, given variables satisfying (III) we have constructed variables satisfying (II) which achieve a greater objective value. This proves that $\bar{\nu} \leq \nu$. This proves the theorem □

Thus, we have shown that the problem (7.7) for a square plant is equivalent to the finite dimensional quadratic programming problem of Theorem 7.2.2 with the dimension known *a priori*. Such types of programming are well studied in the literature and efficient numerical methods are available (e.g., [22]). We should point out that the sum $\sum_{t=0}^{\infty} -\frac{1}{4\bar{c}_{pq}} Z_{pq}(t)^2$ appearing in the quadratic program above is a quadratic function of $y_{ijk\lambda_0}$ with coefficients of the form $< F_{pq}^{ijk\lambda_0}, F_{pq}^{rst\bar{\lambda}_0} >$, which can be readily computed.

The solution procedure consists of solving the quadratic program of Theorem 7.2.2 to obtain the optimal variables $y^0_{ijk\lambda_0}$ and $\Phi^0_{pq}(t)$ with $t = 0, \ldots, L^*$ for all $(p,q) \in MN$. The latter set completely determines Φ^0_{pq} for all $(p,q) \in MN$. From $y^0_{ijk\lambda_0}$ the optimal Φ^0_{pq} for $(p,q) \in M$ can be computed as $\Phi^0_{pq} = \frac{1}{2\bar{c}_{pq}} Z^0_{pq}$ (see Lemma 7.2.3). The quadratic program of Theorem 7.2.2 does not yield immediately any information on Φ^0_{pq} for $(p,q) \in N$. Nontheless, Φ^0_{pq} for $(p,q) \in N$ can be easily obtained once Φ^0_{pq} for $(p,q) \in MN \cup M$ are found through the following (finite dimensional) optimization:

$$minimize \sum_{(p,q)\in N} c_{pq} ||\Phi_{pq}||_1$$

subject to

$$\Phi_{pq} \in R^{L^*}$$

$$\sum_{(p,q)\in N} < F^{ijk\lambda_0}, \Phi_{pq} >= b^{ijk\lambda_0} - \sum_{(p,q)\in(MN)\cup M} < F^{ijk\lambda_0}, \Phi^0_{pq} >$$

This problem can be readily solved via linear programming [3].

From the developments above it follows that the structure of an optimal solution Φ^0 to the primal problem (7.7) has in general an infinite impulse response (IIR). The parts of Φ^0 however that are contained in the cost via their ℓ_1 norm i.e., the Φ^0_{pq}'s with (p,q) in $MN \cup N$, are always FIR.

Finally, it should be noted that the optimal solution has certain properties related to the notion of Pareto optimality (e.g., [22]). In particular, from the uniqueness properties of Φ^0 it is clear that there is no other feasible Φ such that $||\Phi_{pq}||_2 < ||\Phi^0_{pq}||_2$ for some $(p,q) \in MN \cup M$ while $||\Phi_{pq}||_1 \leq ||\Phi^0_{pq}||_1$; or, conversely, there is no Φ such that $||\Phi_{pq}||_1 < ||\Phi^0_{pq}||_1$ for some $(p,q) \in MN \cup M$ while $||\Phi_{pq}||_2 \leq ||\Phi^0_{pq}||_2$.

7.3 The Mixed Problem

In this section we make the statement for the mixed problem precise. We solve the mixed problem via a related problem called the approximate problem. For both the mixed and the approximate problems the following notation is relevant: Let $N_w := \{1, \ldots, n_w\}$ and let $N_z := \{1, \ldots, n_z\}$. Let $\mathcal{S}$ be a given subset of N_z. $\mathcal{S}$ corresponds to those rows of the closed loop which have some part constrained in the ℓ_1 norm. We denote the cardinality of $\mathcal{S}$ by c_n. Let N_p for $p \in \mathcal{S}$ be a subset of N_w. N_p characterizes the part of the pth row of the closed loop that is constrained in the ℓ_1 norm. The (positive) scalars γ_p for $p \in \mathcal{S}$ represent the ℓ_1 constraint level on the pth row. It is assumed that $\gamma_p > \nu_{0,1}$. Finally, $\gamma \in R^{c_n}$ is a vector which has γ_p for $p \in \mathcal{S}$ as its elements.

We define a set $\Gamma_\gamma \subset \ell_1^{n_z \times n_w}$ of feasible solutions as follows: $\Phi \in \ell_1^{n_z \times n_w}$ is in Γ_γ if and only if it satisfies the following conditions:

a) $\sum_{q \in N_p} ||\Phi_{pq}||_1 \leq \gamma_p$ for all $p \in \mathcal{S}$,

b) $\Phi \in \Theta$ (i.e Φ is an achievable closed loop map).

Φ is said to be *feasible* if $\Phi \in \Gamma_\gamma$. Let $\underline{M}$ be a given subset of $N_z \times N_w$.

The problem statements for the mixed and the approximate problems are now presented.

Given a plant G the mixed problem is the following optimization:

$$\mu_\gamma := \inf_{\Phi \in \Gamma_\gamma} \{ \sum_{(p,q) \in \overline{M}} ||\Phi_{pq}||_2^2 \}. \tag{7.11}$$

Given a plant G the approximate problem of order δ is the following optimization:

$$\mu_\gamma^\delta := \inf_{\Phi \in \Gamma_\gamma} \{ \sum_{(p,q) \in \overline{M}} ||\Phi_{pq}||_2^2 + \delta \sum_{p \in \mathcal{S}} \sum_{q \in N_p} ||\Phi_{pq}||_1 \}. \tag{7.12}$$

We will further assume that for all $(p,q) \in N_z \times N_w$ the component Φ_{pq} appears in the ℓ_1 constraint or in the objective function or in both. Note that

$\underline{M}$ is the set of transfer function pairs whose two norms have to be minimized in the problem. The problem is set up so that one can include the constraint of a complete row in the closed loop map Φ or part of a row. This way we can easily incorporate constraints of the form $||\Phi||_1 \leq 1$ which is equivalent to each row having one norm less than 1. Also, the $\mathcal{H}_2$ norm of Φ can be included in the cost as a special case.

We also define the following sets which help in isolating various cases in the dual formulation:

$$\underline{N} := \cup_{i \in \mathcal{S}}(i, N_i),$$

which is set of indices (i, j) such that Φ_{ij} occur in the ℓ_1 constraint,

$$MN := \underline{M} \cap \underline{N},$$

which is the set of indices (i, j) such that Φ_{ij} occurs in the ℓ_1 constraint and its two norm appears in the objective,

$$M := \underline{M} \backslash MN,$$

which is the set of indices (i, j) such that two norm of Φ_{ij} occurs in the objective but it does not appear in the ℓ_1 constraint and

$$N := \underline{N} \backslash MN,$$

which is the set of indices (i, j) such that Φ_{ij} occurs in the ℓ_1 constraint but its two norm does not appear in the objective. With this notation we have, $\underline{M} = (MN) \cup M$ and $\underline{N} = (MN) \cup N$. We assume that $MN \cup M \cup N$ equals $N_z \times N_w$. This implies that for all $(p, q) \in N_z \times N_w$ Φ_{pq} appears in the ℓ_1 constraint or in the objective function or in both.

We define $f_m : \ell_1^{n_z \times n_w} \to R$ and $f_a^\delta : \ell_1^{n_z \times n_w} \to R$ by

$$f_m(\Phi) := \sum_{(p,q) \in \overline{M}} ||\Phi_{pq}||_2^2 = \sum_{(p,q) \in (MN) \cup M} ||\Phi_{pq}||_2^2,$$

and

$$\begin{aligned} f_a^\delta(\Phi) &= \sum_{(p,q) \in \overline{M}} ||\Phi_{pq}||_2^2 + \delta \sum_{p \in \mathcal{S}} \sum_{q \in N_p} ||\Phi_{pq}||_1 \\ &= \sum_{(p,q) \in MN \cup M} ||\Phi_{pq}||_2^2 + \delta \sum_{(p,q) \in MN \cup N} ||\Phi_{pq}||_1, \end{aligned}$$

which are the objective functions of the mixed and the approximate problems respectively. We make the following assumption.

Assumption 6 *The plant is square i.e.,* $n_z = n_u$ *and* $n_y = n_w$.

We now solve the approximate problem and later we give the relation of the mixed problem to the approximate problem.

7.3.1 The Approximate Problem

In this subsection we study the approximate problem of order δ. This problem is very similar to the combination problem. The techniques used in solving the combination problem are often identical to the ones used in solving the approximate problem. We state many facts without proof. These facts can be easily deduced in ways similar to the ones used in the solution of the combination problem. The importance of this problem comes from its connection to the mixed problem.

As in the combination problem, we define for notational convenience

$$Z_{pq}(t) := \sum_{i,j,k,\lambda_0} y_{ijk\lambda_0} F_{pq}^{ijk\lambda_0}(t).$$

Theorem 7.3.1. *There exists $\Phi^0 \in \Gamma_\gamma$ such that*

$$\mu_\gamma^\delta = \sum_{(p,q)\in MN\cup M} \|\Phi_{pq}^0\|_2^2 + \sum_{(p,q)\in MN\cup N} \delta\|\Phi_{pq}^0\|_1.$$

Therefore, the infimum in (7.12) is a minimum. Moreover, the following it is true that μ_γ^δ equals

$$\max \sum_{(p,q)\in M}\sum_{t=0}^{\infty} -\Phi_{pq}(t)^2 + \sum_{(p,q)\in MN}\sum_{t=0}^{\infty} -\Phi_{pq}(t)^2 + \sum_{i,j,k,\lambda_0} y_{ijk\lambda_0} b^{ijk\lambda_0} - \sum_{p\in S} \overline{y}_p \gamma_p,$$

subject to

$$\overline{y} \in R^{c_n},\ \overline{y} \geq 0,\ \ y \in R^{c_s}, \Phi_{pq} \in \ell_1 \ for\ all\ (p,q) \in MN \cup M,$$

$$\left.\begin{array}{lll} -(\delta+\overline{y}_p) \leq |Z_{pq}(t)| \leq (\delta+\overline{y}_p) & if\ (p,q) \in N & \\ 2\Phi_{pq}(t) = Z_{pq}(t) - (\delta+\overline{y}_p) & if\ (p,q) \in MN, & Z_{pq}(t) > (\delta+\overline{y}_p), \\ \quad = Z_{pq}(t) + (\delta+\overline{y}_p) & if\ (p,q) \in MN, & Z_{pq}(t) < -(\delta+\overline{y}_p), \\ \quad = 0 & if\ (p,q) \in MN, & |Z_{pq}(t)| \leq (\delta+\overline{y}_p), \\ \quad = Z_{pq}(t) & if\ (p,q) \in M, & \\ & for\ all\ t = 0,1,2,\ldots. & \end{array}\right\} (IV)$$

In addition, the optimal Φ_{pq}^0 is unique for all $(p,q) \in (MN) \cup M$.

Proof. The proof follows by utilizing results analogous to Lemmas 2 and 3, and similar arguments to Theorem 7.2.1. □

To get an analogous result to Lemma 7.2.3 it is clear that we have to get an *a priori* bound on the dual variable $\overline{y}$.

Lemma 7.3.1. *Let $\Phi^{0,1}$ denote a solution of the standard ℓ_1 problem (7.5). $f_a^\delta(\Phi^{0,1})$ is the objective of the approximate problem evaluated at a solution of the standard ℓ_1 problem. If $(\overline{y}^\gamma,\ y^\gamma)$ is the solution to the approximate problem as given in Theorem 7.3.1 then*

$$\overline{y}_p^\gamma \leq \frac{f_a^\delta(\Phi^{0,1})}{\gamma_p - \nu_{0,1}} \quad for\ all\ p \in S.$$

Proof. Take any $c \in R$ such that

$$\nu_{0,1} < c < \min_{p \in \mathcal{S}} \gamma_p.$$

Let $\gamma^0 \in R^{c_n}$ be given by $\gamma_p^0 = c$. Let

$$\mu_{\gamma^0}^\delta := \inf_{\Phi \in \Gamma_{\gamma^0}} f_a^\delta(\Phi).$$

Then from Corollary 3.3.1 we have,

$$< \gamma - \gamma^0, \overline{y}^\gamma > \leq \mu_{\gamma^0}^\delta - \mu_\gamma^\delta \leq \mu_{\gamma^0}^\delta \leq f_a^\delta(\Phi^{0,1}).$$

Therefore,

$$\sum_{p \in \mathcal{S}} (\gamma_p - c)\overline{y}_p^\gamma \leq f_a^\delta(\Phi^{0,1}).$$

As $(\gamma_p - c) > 0$ we have

$$\overline{y}_p^\gamma \leq \frac{f_a^\delta(\Phi^{0,1})}{\gamma_p - c} \text{ for all } p \in S.$$

This holds for all $c > \nu_{0,1}$ and therefore the lemma follows. □

Now we state the lemma analogous to Lemma 7.2.3.

Lemma 7.3.2. *Let Φ^0 be a solution to the primal problem (7.12) and let $\overline{y}^0$, y^0, Z_{pq}^0 be solutions to the dual. Then the following is true:*

$$\begin{aligned}
-(\delta + \overline{y}_p) \leq Z_{pq}^0(t) &\leq (\delta + \overline{y}_p) && \textit{if } (p,q) \in N, \\
\Phi_{pq}^0(t) &= 0 && \textit{if } (p,q) \in N \textit{ and } |Z_{pq}^0(t)| < (\delta + \overline{y}_p), \\
2\Phi_{pq}^0(t) &= Z_{pq}^0(t) - (\delta + \overline{y}_p) && \textit{if } (p,q) \in MN \textit{ and } Z_{pq}^0(t) > (\delta + \overline{y}_p), \\
&= Z_{pq}^0(t) + (\delta + \overline{y}_p) && \textit{if } (p,q) \in MN \textit{ and } Z_{pq}^0(t) < -(\delta + \overline{y}_p), \\
&= 0 && \textit{if } (p,q) \in MN \textit{ and } |Z_{pq}^0(t)| \leq (\delta + \overline{y}_p), \\
&= Z_{pq}^0(t) && \textit{if } (p,q) \in M.
\end{aligned}$$

Φ_{pq}^0 is unique for all $(p,q) \in (MN) \cup M$. Also, there exists an a priori bound α_a such that $||Z_{pq}^0||_\infty \leq \alpha_a$ where

$$\alpha_a := \frac{f_a^\delta(\Phi^{0,1})}{\gamma_p - \nu_{0,1}} + \delta + \frac{2}{\delta} f_a^\delta(\Phi^{0,1}) + 2\sqrt{f_a^\delta(\Phi^{0,1})}.$$

for all $(p,q) \in N_z \times N_w$. $\Phi^{0,1}$ is a solution to (7.5).

Proof. Similar to the proof of Lemma 7.2.3. Note here that a feasible solution to the approximate problem is always the ℓ_1 optimal $\Phi^{0,1}$ as opposed to H (used in Lemma 7.2.3) which may not be feasible in this case since it may not satisfy the ℓ_1 constraint. This is why $\Phi^{0,1}$ appears in the expression for α_a □

Using arguments similar to the ones used in Lemma 7.2.4 we can determine L_a^* such that

$$|Z_{pq}(t)| < \delta \text{ for all } t \geq L_a^*.$$

The following theorem follows using arguments identical to that used in proving Theorem 7.2.2.

Theorem 7.3.2. *It is true that* μ_γ^δ *equals*

$$\mu_\gamma^\delta = \max\{ \sum_{(p,q)\in M} \sum_{t=0}^{\infty} -\frac{1}{4} Z_{pq}(t)^2 + \sum_{(p,q)\in MN} \sum_{t=0}^{L_a^*} -\Phi_{pq}(t)^2 + \sum_{i,j,k,\lambda_0} y_{ijk\lambda_0} b^{ijk\lambda_0} - \sum_{p\in S} \overline{y}_p \gamma_p \}$$

subject to

$$\begin{array}{ll} \overline{y} \in R^{c_n},\ \overline{y} \geq 0,\ y \in R^{c_s}, \Phi_{pq}(t) \in R^{L_a^*} & \textit{for all } (p,q) \in MN, \\ -(\delta + \overline{y}_p) \leq Z_{pq}(t) \leq (\delta + \overline{y}_p) & \textit{if } (p,q) \in N \\ -(\delta + \overline{y}_p) \leq 2\Phi_{pq}(t) - Z_{pq}(t) \leq (\delta + \overline{y}_p) & \textit{if } (p,q) \in MN \\ \textit{for all } t = 0, 1, 2, \ldots, L_a^*. & \end{array}$$

Furthermore, the optimal Φ_{pq}^0 *of the primal (7.12) is unique for all* $(p,q) \in (MN) \cup M$.

Thus, we have reduced the approximate problem to a finite dimensional quadratic optimization problem with *a priori* known dimension. The same remarks relative to the solution procedure hold as in the combination problem. Note that again for the optimal solution Φ^0 the Φ_{pq}^0's with (p,q) in $MN \cup N$, are always FIR while the Φ_{pq}^0's with (p,q) in M, are IIR.

7.3.2 Relation between the Approximate and the Mixed Problem

In this section we show how to solve the mixed problem using the results of the approximate problem. Note that the approximate problem reduces to a finite dimensional quadratic optimization problem with *a priori* known dimension.

For the mixed problem (1-block) a similar Lagrange duality approach can be used to show that the problem can be converted to a finite dimensional convex problem with some of the optimal Φ_{pq}^0 being possibly FIR and some IIR as in the approximate problem (see Theorem 7.3.1). Nonetheless, even in the single input-single-output-case, an *a priori* bound on the dimension of the equivalent quadratic problem has proved elusive [14]. In addition, the MIMO problem is substantially more complex, for one cannot determine *a priori* which of the optimal dual variables $\overline{y}_p^0$ corresponding to the ℓ_1 constraint is active (i.e., $\overline{y}_p^0 > 0$.) Hence, the *a priori* determination of which (if any) of the otpimal Φ_{pq}^0 is FIR is not possible. This can make the solution procedure extremely complicated and virtually intractable by trying to examine all possibilities.

This difficulty can be circumvented by considering the approximate problem. The results in this section show that a suboptimal solution to the mixed problem can be obtained by solving an approximate problem. The following theorem shows that we can design a controller K for the mixed problem which achieves an objective value within any given tolerance of the optimal value by solving a corresponding approximate problem. The existence of a solution for the mixed problem and the optimal Φ^0_{pq} being unique for $(p,q) \in (MN) \cup M$ can be proved in a similar manner as was done for the approximate problem.

Theorem 7.3.3. $\mu_\gamma \leq \mu^\delta_\gamma \leq \mu_\gamma + \delta|\gamma|_1.$

Proof. It is easy to show that $\mu_\gamma \leq \mu^\delta_\gamma$. Note that if $\Phi \in \Gamma_\gamma$ then

$$\sum_{q \in N_p} \|\Phi_{pq}\|_1 \leq \gamma_p \text{ for all } p \in S.$$

This implies that

$$f^\delta_a(\Phi) \leq \sum_{(p,q)\in(MN)\cup M} \|\Phi_{pq}\|_2^{\ 2} + \delta|\gamma|_1.$$

Taking infimum over Γ_γ on both sides in the above inequality the theorem follows. □

The next theorem is a result on the convergence of the optimal solutions of the approximate problems to the solution of the mixed problem.

Theorem 7.3.4. *Let Φ^n be a solution of the approximate problem of order $\frac{1}{n}$. Then, there exists a subsequence $\{\Phi^{n_k}\}$ of Φ^n and $\Phi^0 \in \ell_1^{n_z \times n_w}$ such that Φ^0 is a solution of the mixed problem and*

$$\Phi^{n_k} \to \Phi^0 \text{ in the } W((\ell_2^{n_z \times n_w})^*, \ell_2^{n_z \times n_w}) \text{ topology.}$$

Furthermore,

$$\Phi^n_{pq} \to \Phi^0 \text{ in the } W((\ell_2)^*, \ell_2) \text{ topology for all } (p,q) \in (MN) \cup M.$$

Proof. See Appendix

7.4 An Illustrative Example

Here we illustrate the theory developed with an example. Consider, the two input single output plant P as depicted in Figure 2 where $u := \begin{pmatrix} u_1 \\ u_2 \end{pmatrix}$ is the input, w is the exogenous disturbance, and y is the measured output. The plant P is given by

$$\hat{P} = (\lambda - 0.5 \;\; 1).$$

The regulated output is given by $z := (y \; u_2)'$. Therefore,

$$\begin{pmatrix} \hat{z} \\ \hat{y} \end{pmatrix} := \begin{pmatrix} \hat{y} \\ \hat{u}_2 \\ \hat{y} \end{pmatrix} = \begin{pmatrix} 1 & \lambda - 0.5 & 1 \\ 0 & 0 & 1 \\ 1 & \lambda - 0.5 & 1 \end{pmatrix} \begin{pmatrix} \hat{w} \\ \hat{u}_1 \\ \hat{u}_2 \end{pmatrix} =: \begin{pmatrix} \hat{G}_{11} & \hat{G}_{12} \\ \hat{G}_{21} & \hat{G}_{22} \end{pmatrix} \begin{pmatrix} \hat{w} \\ \hat{u} \end{pmatrix}.$$

As the plant P is stable a valid Youla parametrization is given by:

$$\hat{H}(\lambda) = \hat{G}_{11} = \begin{pmatrix} 1 \\ 0 \end{pmatrix}, \; \hat{U}(\lambda) = \hat{G}_{12} = \begin{pmatrix} \lambda - 0.5 & 1 \\ 0 & 1 \end{pmatrix} \text{ and } \hat{V}(\lambda) = \hat{G}_{21} = 1.$$

For this problem $n_z = 2$, $n_w = 1$, $n_y = 1$ and $n_u = 2$. Let Φ be an achievable closed loop map then

$$\hat{\Phi}(\lambda) = \hat{H}(\lambda) - \hat{U}(\lambda)\hat{Q}(\lambda)\hat{V}(\lambda).$$

This implies that

$$\begin{pmatrix} \hat{q}_1 \\ \hat{q}_2 \end{pmatrix} = \begin{pmatrix} \frac{1}{\lambda - 0.5} & \frac{-1}{\lambda - 0.5} \\ 0 & 1 \end{pmatrix} \begin{pmatrix} (1 - \hat{\Phi}_1) \\ -\hat{\Phi}_2 \end{pmatrix}.$$

Therefore, q_1 and q_2 are in ℓ_1 only if

$$\frac{1}{\lambda - 0.5}(1 - \hat{\Phi}_1(\lambda) + \hat{\Phi}_2(\lambda)) \in \ell_1.$$

Therefore Φ is an achievable closed loop map if and only if

$$\Phi \in \ell_1^{2\times 1} \text{ and } 1 - \hat{\Phi}_1(0.5) + \hat{\Phi}_2(0.5) = 0.$$

The above equation is the only interpolation condition. Following, the notation developed in the earlier sections we define

$$F_1 := (1, \frac{1}{2}, (\frac{1}{2})^2, \ldots)' \text{ and } F_2 := (-1, -\frac{1}{2}, -(\frac{1}{2})^2)'.$$

It can be checked that the interpolation condition is equivalent to

$$< F, \Phi > = < F_1, \Phi_1 > + < F_2, \Phi_2 > = 1.$$

As $n_u = n_z$ and $n_w = n_y$ the system is square and rank interpolation conditions are absent. First we solve the standard multiple input multiple output ℓ_1 problem for the given system G.

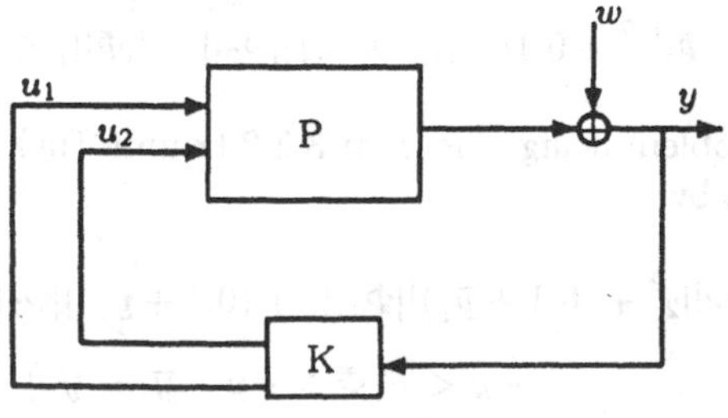

Fig. 7.1. A two input single output example

7.4.1 Standard ℓ_1 Solution

In this subsection we are interested in solving the following optimization:

$$\nu_{0,1} = \inf_{\Phi \text{ Achievable}} ||\Phi||_1 = \inf_{<F,\Phi>=1} ||\Phi||_1.$$

We refer the reader to section 12.1.2 of [3] for the the theory used to solve this problem. It can be easily verified that the above problem reduces to the following finite dimensional linear program:

$$\min_{\nu,\psi,\Phi_i^+,\Phi_i^-} \nu,$$

subject to

$$\begin{aligned} \psi(i) + \sum_{t=0}^{1} \Phi_i^+(t) + \Phi_i^-(t) = \nu \ for\ i = 1,2, \\ \Phi_1^+(0) - \Phi_1^-(0) + \tfrac{1}{2}(\Phi_1^+(1) - \Phi_1^-(1)) - (\Phi_2^+(0) - \Phi_2^-(0)) \\ -\tfrac{1}{2}(\Phi_2^+(1) - \Phi_2^-(1)) = 1, \\ \psi, \Phi_i^+, \Phi_i^- \geq 0. \end{aligned}$$

Using the linear programming software of MATLAB we obtain that an optimal is given by

$$\Phi^{0,1} = \begin{pmatrix} 0.5 \\ -0.5 \end{pmatrix}.$$

This implies that $\nu_{0,1} = 0.5$.

7.4.2 Solution of the Mixed Problem

In this subsection we are interested in solving the following optimization for the given system in Figure 2:

$$\mu_\gamma := \inf_{\Phi \text{ Achievable}} \{||\Phi||_2{}^2 : ||\Phi||_1 \leq 1,\ \Phi \in \ell_1^{2\times 1}\}.$$

We give the corresponding approximate problem of order 0.1 by the the following optimization:

$$\mu_\gamma^{0.1} := \inf_{\Phi \text{ Achievable}} \{||\Phi||_2{}^2 + 0.1||\Phi_1||_1 + 0.1||\Phi_2||_1 : ||\Phi||_1 \leq 1,\ \Phi \in \ell_1^{2\times 1}\}.$$

The dual of the above problem using Theorem 3.3.2 (Kuhn-Tucker-Lagrange duality theorem) is given by:

$$\begin{aligned} \mu_\gamma^{0.1} := \max \inf_{\Phi\in\ell_1^{2\times 1}} \{||\Phi||_2{}^2 + (0.1+\overline{y}_1)||\Phi_1||_1 + (0.1+\overline{y}_2)||\Phi_2||_1 \\ -y < F,\Phi > +y - \overline{y}_1 - \overline{y}_2\} \end{aligned}$$

subject to

$$\overline{y}_1 \geq 0,\ \overline{y}_2 \geq 0, y \in R.$$

In keeping with the notation defined in earlier section we denote

$$Z := yF, \text{ that is } Z_1 = yF_1 \text{ and } Z_2 = yF_2,$$

$$f_a^{0.1}(\Phi) := ||\Phi_1||_2{}^2 + ||\Phi_1||_2{}^2 + 0.1||\Phi_1||_1 + 0.1||\Phi_2||_1.$$

Therefore, we have, $f_a^{0.1}(H) = 1^2 + 0 + 0.1 + 0 = 1.1$ and $f_a^{0.1}(\Phi^{0,1}) = (0.5)^2 + (0.5)^2 + 0.1(0.5+0.5) = 0.6$. Let $\overline{y}^\gamma$ and y^γ be the solution to the dual problem stated above and let Φ^γ be the solution to the primal. We define $L : \ell_1^{2\times 1} \to R$ by $L(\phi) = ||\Phi||_2{}^2 + +(0.1 + \overline{y}_1^\gamma)||\Phi_1||_1 + (0.1 + \overline{y}_2^\gamma)||\Phi_2||_1 - < Z_1^\gamma, \Phi_1 > - < Z_2^\gamma, \Phi_2 >$. From Theorem 3.3.2 (Kuhn-Tucker-Lagrange duality theorem) it follows that Φ^γ minimizes $L(\Phi)$ over all $\Phi \in \ell_1^{2\times 1}$. This implies that $\Phi^\gamma(t)$ minimizes,

$$\Phi_1(t)^2 + (0.1 + \overline{y}_1^\gamma)|\Phi_1(t)| - Z_1^\gamma(t)\Phi_1(t), \tag{7.13}$$

over all $\Phi_1(t) \in R$. We can discard the $\Phi_1(t) \in R$ which have an opposite sign to that of $Z_1^\gamma(t)$ because then $- < Z_1^\gamma, \Phi_1 > \geq 0$. Therefore $\Phi^\gamma(t)$ minimizes,

$$\Phi_1(t)^2 + \Phi_1(t)((0.1 + \overline{y}_1^\gamma)sgn(Z_1^\gamma(t)) - Z_1^\gamma(t)),$$

over all $\Phi_1(t) \in R$ which satisfy $\Phi_1(t)Z_1^\gamma(t) \geq 0$. Without loss of generality assume that $Z_1^\gamma(t) \geq 0$. Then $\Phi^\gamma(t)$ minimizes,

$$\Phi_1(t)^2 + \Phi_1(t)((0.1 + \overline{y}_1^\gamma) - Z_1^\gamma(t)),$$

over all $\Phi_1(t) \in R$ which are positive. Now if $(0.1+\overline{y}_1^\gamma) \geq Z_1^\gamma(t)$ then objective is always positive as $\Phi_1(t)$ is restrained to be positive and therefore the minimizer $\Phi^\gamma(t)$ is forced to be equal to zero. If $0 < (0.1 + \overline{y}_1^\gamma) < Z_1^\gamma(t)$ then the coefficient of $\Phi(t)$ in the objective is negative and therefore we can do better than achieving a zero objective value and this forces $\Phi^\gamma(t) > 0$. With this knowledge we can now differentiate the unconstrained objective function in (7.13) to get $2\Phi^\gamma(t) = Z_1^\gamma(t) - (0.1+\overline{y}_1^\gamma)$. Similarly, if $Z_1^\gamma(t) < -(0.1+\overline{y}_1^\gamma) < 0$ then $2\Phi^\gamma(t) = Z_1^\gamma(t) + (0.1 + \overline{y}_1^\gamma)$. In any case the following holds:

$$|Z_1^\gamma(t)| \leq 0.1 + \overline{y}_1^\gamma + 2|\Phi^\gamma(t)| \leq 0.1 + \frac{f_a^{0.1}(\Phi^{0,1})}{\gamma_p - \nu_{0,1}} + 2||\Phi||_1.$$

The second inequality follows from Lemma 7.3.1. From the fact that $||\Phi_1^\gamma||_1 \leq 1$ we have

$$|Z_1^\gamma(t)| \leq 0.1 + \frac{0.6}{1 - 0.5} + 2 = 3.3.$$

Note that $Z_1^\gamma(t) = y^\gamma F_1(t)$. As $|Z_1^\gamma(0)| \leq 3.3$ it follows that $|y^\gamma F_1(0)| = |y^\gamma| \leq 3.3$. Now,

$$|Z_1^\gamma(t)| = |y^\gamma F_1(t)| \leq 3.3\ F_1(t) = 3.3\frac{1}{2^t}.$$

This implies that we can determine *a priori* L_a^* such that if $t \geq L_a^*$ then

$$|Z_1^\gamma(t)| \leq 0.1 \leq 0.1 + \overline{y}_1^\gamma,$$

which will imply that $\Phi_1^\gamma(t) = 0$ if $t \geq L_a^*$. $L_a^* = 5$ does satisfy this requirement. A similar development holds for Φ_2^γ. From Theorem 7.3.2 we have that the dual can be written as:

$$\begin{aligned} \mu_\gamma^{0.1} = \quad & \max\{\sum_{i=1}^{2}\sum_{t=0}^{5} -\Phi_i(t)^2 + y - \sum_{i=1}^{2}\overline{y}_i\} \\ \textit{subject to} \quad & \\ & \overline{y} \in R^2,\ \overline{y} \geq 0\ y \in R, \Phi_i \in R^6\ i = 1, 2 \\ & -(0.1 + \overline{y}_i) \leq 2\Phi_i(t) - Z_i(t) \leq (0.1 + \overline{y}_i) \text{ for } i = 1, 2. \\ & \text{for all } t = 0, 1, 2, \ldots, 6. \end{aligned}$$

Using MATLAB software we obtain that the optimal Φ^γ which is unique for this example is given by

$$\hat{\Phi}_1^\gamma(\lambda) = 0.3972 + 0.1732\lambda + 0.0617(\lambda)^2 + 0.0058(\lambda)^3,$$

$$\hat{\Phi}_2^\gamma(\lambda) = -\Phi_1^\gamma(\lambda).$$

Therefore, $f_a^{0.1}(\gamma) = 0.5109$, $||\Phi^\gamma||_1 = 0.6379$ and $||\Phi^\gamma||_2 = 0.6191$. This implies from Theorem 7.3.3 that if Φ^0 represents the solution of the mixed problem then

$$|f_m(\Phi^0) - f_m(\Phi^\gamma)| \leq 0.2.$$

This completes the example.

7.5 Summary

In this chapter we considered two related problems of MIMO controller design which incorporate the $\mathcal{H}_2$ and the ℓ_1 norms of input-output maps constituting the closed loop directly in their definitions.

In the first problem termed as the combination problem, a positive linear combination of the square of the $\mathcal{H}_2$ norms and the ℓ_1 norms of the input-output maps was minimized over all stabilizing controllers. It was shown that, for the 1-block case, the optimal is possibly IIR and the solution can be nonunique. However, it was shown that the problem can be solved exactly via a finite dimensional quadratic optimization problem and a linear programming problem of *a priori* known dimensions.

In the second problem termed as the mixed problem, the $\mathcal{H}_2$ performance of the closed loop is minimized subject to a ℓ_1 constraint. It was shown that suboptimal solutions within any given tolerance of the optimal value can be obtained via the solution to a related combination problem.

7.6 Appendix

7.6.1 Interpolation Conditions

We analyse here in some detail the zero interpolation conditions.

Given $\alpha_i \in \ell_1^{1\times n_s}$ and $\beta_j \in \ell_1^{n_w \times 1}$ we have

$$\begin{aligned}
\hat{\alpha}_i(\lambda)\hat{\Phi}(\lambda)\hat{\beta}_j(\lambda) &= \sum_{p=1}^{n_s}\sum_{q=1}^{n_w} \hat{\alpha}_{ip}(\lambda)\hat{\Phi}_{pq}(\lambda)\hat{\beta}_{jq}(\lambda) \\
&= \sum_{p=1}^{n_s}\sum_{q=1}^{n_w}\sum_{t=0}^{\infty} (\alpha_{ip} * \Phi_{pq} * \beta_{jq})(t)\lambda^t \\
&= \sum_{p=1}^{n_s}\sum_{q=1}^{n_w}\sum_{t=0}^{\infty}\sum_{l=0}^{\infty} \alpha_{ip}(t-l)(\Phi_{pq} * \beta_{jq})(l)\lambda^t \\
&= \sum_{p=1}^{n_s}\sum_{q=1}^{n_w}\sum_{t=0}^{\infty}\sum_{l=0}^{\infty} \alpha_{ip}(t-l)\sum_{s=0}^{\infty}\beta_{jq}(l-s)\Phi_{pq}(s)\lambda^t \\
&= \sum_{p=1}^{n_s}\sum_{q=1}^{n_w}\sum_{t=0}^{\infty}\sum_{l=0}^{\infty}\sum_{s=0}^{\infty} \alpha_{ip}(t-l)\beta_{jq}(l-s)\Phi_{pq}(s)\lambda^t .
\end{aligned}$$

Therefore, it follows that

$$\begin{aligned}
(\hat{\alpha}_i\hat{\Phi}\hat{\beta}_j)^{(k)}(\lambda_0) &= \sum_{p=1}^{n_s}\sum_{q=1}^{n_w}\sum_{t=0}^{\infty}\sum_{l=0}^{\infty}\sum_{s=0}^{\infty} \alpha_{ip}(t-l)\beta_{jq}(l-s)\Phi_{pq}(s)\ (\lambda^t)^{(k)}\Big|_{\lambda=\lambda_0} . \\
&= \sum_{p=1}^{n_s}\sum_{q=1}^{n_w}\sum_{s=0}^{\infty}\sum_{t=0}^{\infty}\sum_{l=0}^{\infty} \alpha_{ip}(t-l)\beta_{jq}(l-s)\Phi_{pq}(s)\ (\lambda^t)^{(k)}\Big|_{\lambda=\lambda_0} .
\end{aligned}$$

Define $F^{ijk\lambda_0} \in \ell_\infty^{n_s\times n_w}$ by

$$F_{pq}^{ijk\lambda_0}(s) := \sum_{l=0}^{\infty}\sum_{t=0}^{\infty} \alpha_{ip}(t-l)\beta_{jq}(l-s)\ (\lambda^t)^{(k)}\Big|_{\lambda=\lambda_0} . \tag{7.14}$$

It can be easily verified that for any $\Phi \in \ell_1^{n_s\times n_w}$

$$(\hat{\alpha}_i\hat{\Phi}\hat{\beta}_j)^{(k)}(\lambda_0) = <\Phi, F^{ijk\lambda_0}> . \tag{7.15}$$

Proof of Lemma 7.1.1 : We first show that if $|\lambda| < 1$ then there exists an integer T such that $t > T$ implies $|(\lambda^t)^k| \le |(\lambda^{t+1})^k|$ where t is an integer. Let T be any integer such that $T > \frac{|\lambda|+k-1}{1-|\lambda|}$ and let $t > T$ be an integer then

$$\begin{aligned}
|(\lambda^t)^{(k)}| - |(\lambda^{t+1})^{(k)}| &= |(t)(t-1),\ldots,(t-k+1)\lambda^{t-k}| \\
&\quad -|(t+1)(t),\ldots,(t-k+2)\lambda^{t-k+1}| \\
&= (t)(t-1),\ldots,(t-k+2)|\lambda|^{t-k}. \\
&\quad \{t-k+1-(t+1)|\lambda|\} \\
&= (t)(t-1),\ldots,(t-k+2)|\lambda|^{t-k}. \\
&\quad \{t(1-|\lambda|)-k-|\lambda|+1\} \\
&= (t)(t-1),\ldots,(t-k+2)|\lambda|^{t-k}(1-|\lambda|). \\
&\quad \{t-(\tfrac{|\lambda|+k-1}{1-|\lambda|})\} \\
&\ge 0.
\end{aligned}$$

Suppose s is an integer such that $s \geq T$ where T is as defined above. From (7.3) we have

$$
\begin{aligned}
|F_{pq}^{ijk\lambda_0}(s)| &= \left| \left[\sum_{l=0}^{\infty} \sum_{t=0}^{\infty} \alpha_{ip}(t-l)\beta_{jq}(l-s)(\lambda^t)^{(k)} \right]_{\lambda=\lambda_0} \right| \\
&\leq \left[\sum_{l=s}^{\infty} \sum_{t=l}^{\infty} |\alpha_{ip}(t-l)|\ |\beta_{jq}(l-s)||(\lambda^t)^{(k)}| \right]_{\lambda=\lambda_0} \\
&\leq \left[\sum_{l=s}^{\infty} \sum_{t=s}^{\infty} |\alpha_{ip}(t-l)|\ |\beta_{jq}(l-s)|\ |(\lambda^t)^{(k)}| \right]_{\lambda=\lambda_0} \\
&\leq \sup_{s \leq t < \infty} |(\lambda^t)^{(k)}|_{\lambda=\lambda_0} \sum_{l=s}^{\infty} \sum_{t=s}^{\infty} |\alpha_{ip}(t-l)|\ |\beta_{jq}(l-s)| \\
&\leq \sup_{s \leq t < \infty} |(\lambda^t)^{(k)}|_{\lambda=\lambda_0} \sum_{l=s}^{\infty} |\beta_{jq}(l-s)| \sum_{t=s}^{\infty} |\alpha_{ip}(t-l) \\
&\leq \sup_{s \leq t < \infty} |(\lambda^t)^{(k)}|_{\lambda=\lambda_0} ||\alpha_{ip}||_1\ ||\beta_{jq}||_1 \\
&= (s)(s-1)\dots(s-k+1)|\lambda_0|^{s-k}||\alpha_{ip}||_1\ ||\beta_{jq}||_1.
\end{aligned}
$$

From the ratio test as $|\lambda_0| < 1$ it follows that $|F_{pq}^{ijk\lambda_0}| \in \ell_1$ because

$$\lim_{s \to \infty} \frac{(s+1)(s)\dots(s-k+2)|\lambda_0|^{(s+1)}}{(s)(s-1)\dots(s-k+1)|\lambda_0|^s} = |\lambda_0| < 1.$$

Note that given any $\epsilon > 0$ we can always choose a $T_0 \geq T$ such that

$$s^k|\lambda_0|^{s-k}||\alpha_{ip}||_1\ ||\beta_{jq}||_1 < \epsilon \text{ for all } s > T_0,$$

which means that we can choose $T_0 \geq T$ such that

$$|F_{pq}^{ijk\lambda_0}(s)| < \epsilon \text{ for all } s > T_0.$$

This proves the lemma. □

The the elements $G_{\alpha_i qt}$ and $G_{\beta_j pt}$ corresponding to the rank conditions can be defined as [3]:

$$G_{\alpha_i qt}(l) := \begin{pmatrix} \vdots \dots \vdots & \overbrace{\vdots}^{q^{th}\ column} & \vdots \dots \vdots \\ 0 \dots 0 & \alpha_i'(t-l) & 0 \dots 0 \\ \vdots \dots \vdots & \vdots & \vdots \dots \vdots \end{pmatrix},$$

$$G_{\beta_j pt}(l) := \begin{pmatrix} \dots & 0 & \dots \\ & \vdots & \\ \dots & 0 & \dots \\ \dots & \beta_j'(t-l) & \dots \\ \dots & 0 & \dots \\ & \vdots & \\ \dots & 0 & \dots \end{pmatrix} \}p^{th}\ row.$$

As $\hat{\alpha}_i$ and $\hat{\beta}_j$ are polynomial vectors we have that $G_{\alpha_i qt}$ and $G_{\beta_j pt}$ are in $\ell_1^{n_z \times n_w}$.

7.6.2 Existence of a Solution for the Combination Problem

Here we show that a solution to (7.7) always exists.
Proof of Lemma 7.2.1 : As

$$f(\Phi) := \sum_{(p,q)\in MN\cup M} \bar{c}_{pq}||\Phi_{pq}||_2^{\,2} + \sum_{(p,q)\in MN\cup N} c_{pq}||\Phi_{pq}||_1,$$

we have

$$\nu_e = \inf_{\Phi\in\Theta_e} \{f(\Phi) : f(\Phi) \le \nu_e + 1\}.$$

This implies that we can restrict Φ in the infimization to satisfy the conditions that $||\Phi_{pq}||_2^{\,2}$ is bounded above by some constant, $\overline{C}$ for all $(p,q) \in M$ and the $||\Phi_{pq}||_1$ is bounded above by some constant, C for all $(p,q) \in MN \cup N$. Define

$$B = \{\Phi \in \ell_2^{n_z \times n_w} : ||\Phi_{pq}||_2^{\,2} \le \overline{C} \text{ for all } (p,q) \in M \text{ and } \\ ||\Phi_{pq}||_1 \le C \text{ for all } (p,q) \in MN \cup N\}.$$

Therefore,

$$\nu_e = \inf_{\Phi\in\Theta_e\cap B} f(\Phi).$$

From, above we conclude that there exists a sequence $\{\Phi^n\} \in \Theta_e \cap B$ such that

$$f(\Phi^n) \le \nu_e + \frac{1}{n}. \tag{7.16}$$

As $\{\Phi^n\} \in \Theta_e \cap B$ we have the following

$$< F^{ijk\lambda_0}, \Phi^n >= b^{ijk\lambda_0}, \tag{7.17}$$

$$||\Phi^n_{pq}||_1 \le C \text{ for all } (p,q) \in MN \cup N, \tag{7.18}$$

$$||\Phi^n_{pq}||_2^{\,2} \le \overline{C} \text{ for all } (p,q) \in M. \tag{7.19}$$

From (7.18) we conclude that for all $(p,q) \in MN \cup N$, Φ^n_{pq} belongs to a bounded set in $(c_0)^*$. From the Banach-Alaoglu result (see Theorem 2.3.1) and separability of c_0 we conclude that for all $(p,q) \in MN \cup N$ there exists a subsequence $\{\Phi^{n_k}_{pq}\}$ of $\{\Phi^n_{pq}\}$ and Φ^0_{pq} such that $\{\Phi^{n_k}_{pq}\} \to \Phi^0_{pq}$ in the $W((c_0)^*, c_0)$ topology. This implies that

$$< v, \Phi^{n_k}_{pq} >\to< v, \Phi^0_{pq} > \text{ for all } v \in c_0. \tag{7.20}$$

Similarly, we conclude that for all $(p,q) \in M$ there exists a subsequence $\{\Phi^{n_{k_s}}_{pq}\}$ of $\{\Phi^{n_k}_{pq}\}$ and Φ^0_{pq} such that $\{\Phi^{n_{k_s}}_{pq}\} \to \Phi^0_{pq}$ in the $W((\ell_2)^*, \ell_2)$ topology. This implies that

$$< v, \Phi_{pq}^{n_{k_s}} > \to < v, \Phi_{pq}^0 > \text{ for all } v \in \ell_2. \tag{7.21}$$

Thus, we have defined $\Phi^0 \in \mathcal{A}$ by the limits (7.20) and (7.21). Note that

$$< F^{ijk\lambda_0}, \Phi > = \sum_{(p,q) \in N_z \times N_w} < F_{pq}^{ijk\lambda_0}, \Phi_{pq} > .$$

Therefore, it follows from (7.17), (7.21), (7.20) and Lemma 7.1.1 that

$$< F^{ijk\lambda_0}, \Phi^0 > = b^{ijk\lambda_0}.$$

Similarly, the rank interpolation conditions are also satisfied by Φ^0. From the above discussion it follows that Φ^0 is in Θ_e and therefore,

$$f(\Phi^0) = \sum_{(p,q) \in (MN) \cup M} \bar{c}_{pq} ||\Phi_{pq}^0||_2^2 + \sum_{(p,q) \in N} c_{pq} ||\Phi_{pq}^0||_1 \geq \nu_e.$$

From (7.20) and (7.21) it follows that

$$\Phi_{pq}^{n_{k_s}}(t) \to \Phi_{pq}^0(t), \text{ for all } t \in R \text{ and for all } (p,q) \in N_z \times N_w.$$

This implies that for all T as $s \to \infty$

$$\sum_{t=0}^{T} \Big(\sum_{(p,q) \in (MN) \cup M} \bar{c}_{pq} |\Phi_{pq}^{n_{k_s}}(t)|_2^2 + \sum_{(p,q) \in (MN) \cup N} c_{pq} |\Phi_{pq}^{n_{k_s}}(t)| \Big) \to$$

$$\sum_{t=0}^{T} \Big(\sum_{(p,q) \in (MN) \cup M} \bar{c}_{pq} |\Phi_{pq}^0(t)|_2^2 + \sum_{(p,q) \in (MN) \cup N} \bar{c}_{pq} |\Phi_{pq}^0(t)|_2^2 \Big)$$

We have from (7.16) that for all s and T

$$\sum_{t=0}^{T} \Big(\sum_{(p,q) \in (MN) \cup M} \bar{c}_{pq} |\Phi_{pq}^{n_{k_s}}(t)|_2^2 + \sum_{(p,q) \in (MN) \cup N} c_{pq} |\Phi_{pq}^{n_{k_s}}(t)| \Big) \leq \nu_e + \frac{1}{n_{k_s}}. \tag{7.22}$$

In (7.22) first letting $s \to \infty$ and then letting $T \to \infty$ we have that $f(\Phi^0) \leq \nu_e$. This proves the lemma. □

7.6.3 Results on the Mixed Problem

Proof of Theorem 7.3.4 : From Theorem 7.3.3 we have that

$$f_{\alpha}^{\frac{1}{n}}(\Phi^n) \leq \mu_\gamma + \frac{1}{n} |\gamma|_1.$$

This implies that there exists a constant C_1 such that $||\Phi_{pq}^n||_2 \leq C_1$, for all $(p,q) \in (MN) \cup M$. From the Banach-Alaoglu (see Theorem 2.3.1) result we conclude that there exists a subsequence $\{\Phi_{pq}^{n_k}\}$ of $\{\Phi_{pq}^n\}$ and $\Phi_{pq}^0 \in \ell_2$ such that

$$\Phi_{pq}^{n_k} \to \Phi_{pq}^0 \text{ in the } W((\ell_2)^*, \ell_2) \text{ topology for all } (p,q) \in (MN) \cup M.$$

This implies that for all $v \in \ell_2$ and for all $(p,q) \in (MN) \cup M$,

$$< v, \Phi^{n_k}_{pq} > \to < v, \Phi^0_{pq} > \text{ as } k \to \infty. \tag{7.23}$$

$\Phi^{n_k} \in \Gamma_\gamma$ for every k therefore

$$\sum_{q \in N_p} ||\Phi^{n_k}_{pq}||_1 \leq \gamma_p \text{ for all } p \in \mathcal{S}.$$

We conclude that there exists a subsequence $\{\Phi^{n_{k_s}}_{pq}\}$ of $\{\Phi^{n_k}_{pq}\}$ and $\overline{\Phi}^0_{pq} \in \ell_1^{n_z \times n_w}$ such that for all $v \in c_0$ and for all $(p,q) \in (MN) \cup N$,

$$< v, \Phi^{n_{k_s}}_{pq} > \to < v, \overline{\Phi}^0_{pq} > \text{ as } s \to \infty. \tag{7.24}$$

From the uniqueness of the limit for all $(p,q) \in ((MN) \cup M) \cap ((MN) \cup N)$, $\overline{\Phi}^0_{pq} = \Phi^0_{pq}$. Thus, for every $(p,q) \in N_z \times N_w$ we have a sequence $\{\Phi^{n_{k_s}}_{pq}\}$ which converges to Φ^0_{pq} in the $W((\ell_2)^*, \ell_2)$ topology (note convergence in $W((c_0)^*, c_0)$ implies convergence in $W((\ell_2)^*, \ell_2)$). For all s we have

$$\sum_{(p,q) \in (MN) \cup M} ||\Phi^{n_{k_s}}_{pq}||_2^2 \leq \mu_\gamma + \frac{1}{n_{k_s}} |\gamma|_1.$$

From convergence in the ℓ_2 weak star topology we have

$$\sum_{(p,q) \in (MN) \cup M} ||\Phi^0_{pq}||_2^2 \leq \mu_\gamma.$$

Similarly, it can be shown that

$$\sum_{q \in N_p} ||\Phi^0_{pq}||_1 \leq \gamma_p \text{ for all } p \in \mathcal{S}.$$

Also, from equations (7.23), (7.24) and the fact that the zero interpolation conditions can be characterized by elements in ℓ_2 we know that Φ^0 satisfies the zero interpolation conditions. To prove that $\Phi^0 \in \Gamma_\gamma$ we still have to prove that $\Phi^0_{pq} \in \ell_1$ for all $(p,q) \in M$. This follows in a similar manner to that of the combination problem. Analogous to what was done in the combination problem we define $\mathcal{A}$ and Θ_e to be the sets

$$\{\Phi \in \ell_2^{n_z \times n_w} : \Phi_{pq} \in \ell_1 \text{ for all } (p,q) \in (MN) \cup N\} \text{ and}$$

$$\{\Phi \in \mathcal{A} : \Phi \text{ satisfies the zero and the rank interpolation conditions}\},$$

respectively. Finally

$$\underline{\Gamma}_\gamma := \{\Phi \in \Theta_e : \sum_{q \in N_p} ||\Phi_{pq}||_1 \leq \gamma_p \text{ for all } p \in \mathcal{S}\}.$$

We also define the corresponding optimization problem μ^e_γ as

$$\mu^e_\gamma := \inf_{\Phi \in \underline{\Gamma}_\gamma} f_m(\Phi). \tag{7.25}$$

We can show in a similar manner to the combination problem that the solution to (7.25) exists and is such that its solution is in Θ. This proves that $\mu_\gamma^e = \mu_\gamma$ and also that the solution to (7.25) is a solution to (7.11). The Φ^0 we have constructed is a solution to (7.25) and therefore we conclude that it is a solution of (7.11) and that

$$\sum_{(p,q)\in(MN)\cup M} \|\Phi_{pq}^0\|_2^2 = \mu_\gamma.$$

The statement about the original sequence converging for $(p,q) \in (MN) \cup M$ follows from the fact that for the mixed problem Φ_{pq}^0 is unique if $(p,q) \in (MN) \cup M$. This proves the theorem. □

8. Multiple-input Multiple-output Systems

Most approaches which incorporate the ℓ_1 objective characterize the achievability of a closed loop map through a stabilizing controller by using zero interpolation conditions on the closed loop map. This was the approach taken in Chapter 7. Computation of the zeros and the zero directions can be done by finding the nullspaces of certain Toeplitz like matrices [3]. Once the optimal closed loop map is determined the task of determining the controller still remains. The closed loop map needs to satisfy the zero interpolation conditions exactly to guarantee that the correct cancellations take place while solving for the controller. However, numerical errors are always present and there exists a need to determine which poles and zeros cancel. These difficulties exist even for the pure MIMO ℓ_1 problem, when zero interpolation methods are employed. However, in [24] it was shown that converging upper and lower bounds can be determined to the ℓ_1 problem by solving an auxiliary problem which does not require zero interpolation and thus avoids the above mentioned problems.

In this chapter we study the $\mathcal{H}_2 - \ell_1$ problem for the general case. Unlike the square case it is difficult to obtain exact solutions in the general case. We show that converging upper and lower bounds can be computed without zero interpolation for the most general MIMO case. This provides an attractive method for solving multi-objective problems which incorporate ℓ_1 and $\mathcal{H}_2$ objectives.

This chapter is organized as follows. In Section 8.1 we formulate the problem and define an auxiliary problem which regularizes the original one. In Section 8.2 we present converging lower and upper bounds for the problem. Finally, we conclude in Section 8.3.

8.1 Problem Statement

Let H, U and V in the Youla parametrization be partitioned into submatrices as given below

$$H = \begin{pmatrix} H^{11} & H^{12} \\ H^{21} & H^{22} \end{pmatrix}, \; U = \begin{pmatrix} U^1 \\ U^2 \end{pmatrix} \text{ and } V = \begin{pmatrix} V^1 & V^2 \end{pmatrix},$$

according to the following equation

$$H - U * Q * V = \begin{pmatrix} H^{11} & H^{12} \\ H^{21} & H^{22} \end{pmatrix} - \begin{pmatrix} U^1 \\ U^2 \end{pmatrix} * Q * (V^1 \; V^2),$$

for some $Q \in \ell_1^{n_u \times n_y}$, where n_u is the number of inputs and n_y is the number of measured outputs. The problem statement is: *Given a plant G, positive real number γ solve the following problem.*

$$\inf_{Q \in \ell_1^{n_u \times n_y}} \|H^{22} - U^2 * Q * V^2\|_2^2$$

subject to

$$\|H^{11} - U^1 * Q * V^1\|_1 \leq \gamma.$$

We denote by μ the optimal value obtained from the above problem.

Now we define an auxiliary problem which is intimately related to the one defined above. The auxiliary problem statement is: *Given a plant G, positive real numbers α and γ solve the following problem.*

$$\begin{aligned} &\inf_{Q \in \ell_1^{n_u \times n_y}} \|H^{22} - U^2 * Q * V^2\|_2^2 \\ &\text{subject to} \\ &\|H^{11} - U^1 * Q * V^1\|_1 \leq \gamma \\ &\|Q\|_1 \leq \alpha. \end{aligned} \tag{8.1}$$

The optimal value obtained from the above problem is denoted by ν.

Note that in the problem statement of μ the allowable Youla parameter Q which is in $\ell_1^{n_u \times n_y}$ needs to satisfy $\|H^{11} - U^1 * Q * V^1\|_1 \leq \gamma$. Therefore it follows that $\|U^1 * Q * V^1\|_1 = \|H^{11} - U^1 * Q * V^1 - H^{11}\|_1 \leq \|H^{11} - U^1 * Q * V^1\|_1 + \|H^{11}\|_1 \leq \|H^{11}\|_1 + \gamma$. Suppose, $\hat{U}^1$ has more rows than columns and $\hat{V}^1$ has more columns than rows and both have full normal rank. Thus the left inverse of $\hat{U}^1$ exists (given by $(\hat{U}^1)^{-l}$) and the right inverse of $\hat{V}^1$ exists (given by $(\hat{V}^1)^{-r}$). Further suppose that $\hat{U}^1$ and $\hat{V}^1$ have no zeros on the unit circle. Then it can be shown that there exists a β (which depends only on $(\hat{U}^1)^{-l}$ and $(\hat{V}^1)^{-r}$) such that $\|Q\|_1 \leq \beta$. Indeed as U^1 and V^1 are left and right invertible it follows that $\hat{Q} = (\hat{U}^1)^{-l}\hat{R}(\hat{V}^1)^{-r}$. This implies that $\|Q\|_1 \leq \|(U^1)^{-l}\|_1(\|H^{11}\|_1 + \gamma)\|(V^1)^{-r}\|_1 =: \beta$. The assumption that $\hat{U}^1$ and $\hat{V}^1$ have no zeros on the unit circle ensures that β is finite. Thus if in the auxiliary problem we choose $\alpha = \beta$ then the constraint $\|Q\|_1 \leq \alpha$ is redundant in the problem statement of ν and we get $\mu = \nu$. The extra constraint in the problem statement of ν is useful because it regularizes the problem (as will be seen).

The following lemma is a result on the uniqueness of the solution to (8.1).

Lemma 8.1.1. *Let Q^0 in $\ell_1^{n_u \times n_y}$ be a solution to (8.1). Let $\Phi^0 = H - U * Q^0 * V$ with $\Phi^{22,o} = H^{22} - U^2 * Q^0 * V^2$ and $\Phi^{11,o} = H^{11} - U^1 * Q^0 * V^1$. Then $\Phi^{22,o}$ is unique. Furthermore, if $\hat{U}^2$ and $\hat{V}^2$ have full normal column and row ranks respectively then Q^0 is unique.*

Proof. Note that the problem statement of ν given by (8.1) can be recast as,

$$\nu = \inf\{||\Phi^{22}||_2^2 : \Phi^{22} \in A_{al}\}, \tag{8.2}$$

where $A_{al} = \{\Phi^{22}$: there exists $Q \in \ell_1^{n_u \times n_y}$ with $\Phi^{22} = H^{22} - U^2 * Q * V^2, ||H^{11} - U^1 * Q * V^1||_1 \leq \gamma,$ and $||Q||_1 \leq \alpha\}$. It is clear that A_{al} is a convex set. It is also true that $||.||_2^2$ is a strictly convex function. It follows from Lemma 3.3.2 that the minimizer of (8.2) given by $\Phi^{22,o}$, if it exists is unique. If $\hat{U}^2$ and $\hat{V}^2$ have full column and row ranks then it follows that

$$\hat{Q}^0 = (\hat{U}^2)^{-l}\hat{\Phi}^{22,o}(\hat{V}^2)^{-r},$$

where $(\hat{U}^2)^{-l}$ and $(\hat{V}^2)^{-r}$ represent the left and the right inverses of $\hat{U}^2$ and $\hat{V}^2$ respectively. Thus $\hat{Q}^0$ is unique. This proves the lemma. □

8.2 Converging Lower and Upper Bounds

In this section we will obtain converging upper and lower bounds to the auxillary problem. The following lemma will be useful towards this goal.

Lemma 8.2.1. *Suppose ϕ_k is a sequence in ℓ_2, ϕ_0 is in ℓ_2 and $\phi_k(t) \to \phi_0(t)$ for all t. Suppose also that $||\phi_k||_2 \nearrow ||\phi_0||_2$. Then $||\phi_k - \phi_0||_2 \to 0$.*

Proof. Given $\epsilon > 0$ choose n such that

$$||(I - P_n)\phi_0||_2^2 \leq \min\{\tfrac{\epsilon}{8}, (\frac{\epsilon}{8(||\phi_0||_2 + 1)})^2\}, \tag{8.3}$$

where P_n is the truncation operator. As $\phi_k(t) \to \phi_0(t)$ we can choose K_2 such that

$$k > K_2 \Rightarrow ||P_n(\phi_k - \phi_0)||_2^2 \leq \tfrac{\epsilon}{4}. \tag{8.4}$$

We know that $||P_n(\phi_k)||_2 \to ||P_n(\phi_0)||_2$ as $k \to \infty$. From above and the fact that $||\phi_k||_2 \to ||\phi_0||_2$ it follows that we can choose K_3 such that

$$k > K_3 \Rightarrow |\; ||(I - P_n)\phi_k||_2^2 - ||(I - P_n)\phi_0||_2^2 \;| \leq \tfrac{\epsilon}{4}. \tag{8.5}$$

Let $K \geq \max\{K_2, K_3\}$ the $k > K$ implies

$$\begin{aligned}
||\phi_k - \phi_0||_2^2 &= ||P_n(\phi_k - \phi_0)||_2^2 + ||(I - P_n)(\phi_k - \phi_0)||_2^2 \\
&\leq \tfrac{\epsilon}{4} + ||(I - P_n)(\phi_k)||_2^2 + ||(I - P_n)(\phi_0)||_2^2 \\
&\qquad +2 \sum_{t=n+1}^{\infty} |\phi_k(t)|\;|\phi_0(t)| \\
&\leq \tfrac{\epsilon}{4} + 2||(I - P_n)(\phi_0)||_2^2 + \tfrac{\epsilon}{4} + 2 \sum_{t=n+1}^{\infty} |\phi_k(t)|\;|\phi_0(t)| \\
&\leq \tfrac{\epsilon}{4} + 2\tfrac{\epsilon}{8} + \tfrac{\epsilon}{4} + 2||(I - P_n)\phi_k||_2 \\
&\quad ||(I - P_n)\phi_0||_2 \\
&\leq \tfrac{\epsilon}{4} + 2\tfrac{\epsilon}{8} + \tfrac{\epsilon}{4} + 2||\phi_0||_2 \frac{\epsilon}{8(||\phi_0||_2 + 1)} \\
&\leq \epsilon.
\end{aligned}$$

□

8.2.1 Converging Lower Bounds

Let ν_n be defined by

$$\begin{aligned} &\inf_{Q\in\ell_1^{n_u\times n_y}} \quad \|P_n(H^{22} - U^2 * Q * V^2)\|_2^2 \\ &\text{subject to} \\ &\qquad \|P_n(H^{11} - U^1 * Q * V^1)\|_1 \leq \gamma \\ &\qquad \|Q\|_1 \leq \alpha. \end{aligned} \tag{8.6}$$

It is clear that only the parameters of $Q(0),\ldots,Q(n)$ enter into the optimization problem and therefore (8.6) is a finite dimensional quadratic programming problem. Once optimal $Q(0),\ldots,Q(n)$ are found, then $Q = \{Q(0),\ldots,Q(n),0,\ldots\}$ will be an FIR optimal solution to (8.6).

Theorem 8.2.1. *Suppose the constraint set in problem (8.1) is nonempty. Then problem (8.1) always has an optimal solution Q^0 in $\ell_1^{n_u\times n_y}$. Furthermore,*

$$\nu_n \nearrow \nu.$$

Also, if $\Phi^{22,o} := H^{22} - U^2 * Q^0 * V^2$ and $\Phi^{22,n} := H^{22} - U^2 * Q^n * V^2$ where Q^n is a solution to (8.6) *then there exists a subsequence $\{\Phi^{22,n_m}\}$ of the sequence $\{\Phi^{22,n}\}$ such that*

$$\|\Phi^{22,n_m} - \Phi^{22,o}\|_2 \to 0 \text{ as } m \to \infty.$$

If $\hat{U}^2$ and $\hat{V}^2$ have full normal column and row ranks respectively then Q^0 is unique and

$$\|\Phi^{22,n} - \Phi^{22,o}\|_2 \to 0 \text{ as } n \to \infty.$$

Proof. We know that for any Q in $\ell_1^{n_u\times n_y}$, $\|P_n(H^{11} - U^1 * Q * V^1)\|_1 \leq \|P_{n+1}(H^{11} - U^1 * Q * V^1)\|_1$ and $\|P_n(H^{22} - U^2 * Q * V^2)\|_2^2 \leq \|P_{n+1}(H^{22} - U^2 * Q * V^2)\|_2^2$. Therefore $\nu_n \leq \nu_{n+1}$ for all $n = 1,2,\ldots$. Thus $\{\nu_n\}$ forms an increasing sequence. Similarly it can be shown that for all n, $\nu_n \leq \nu$.

For $n = 1,2,\ldots$, let $\{Q^n\}$ in $\ell_1^{n_u\times n_y}$ be FIR solutions of (8.6). As the sequence $\{Q^n\}$ is uniformly bounded by α in $\ell_1^{n_u\times n_y}$ it follows from Banach-Alaoglu theorem (see Theorem 2.3.1) that there exists a subsequence $\{Q^{n_m}\}$ of $\{Q^n\}$ and Q^0 in $\ell_1^{n_u\times n_y}$ such that $Q_{ij}^{n_m}$ converges to Q_{ij}^0 in the $W(c_0^*, c_0)$ topology. This implies that $Q^{n_m}(t)$ converges to $Q^0(t)$ for all $t = 0,1,\ldots$. Therefore for all n, $P_n(U * Q^{n_m} * V)$ converges to $P_n(U * Q^0 * V)$ as m tends to ∞. Now for any $n > 0$ and for any $n_m > n$, $\|P_n(H^{11} - U^1 * Q^{n_m} * V^1)\|_1 \leq \gamma$. This implies that $\|P_n(H^{11} - U^1 * Q^0 * V^1)\|_1 \leq \gamma$. Since n is arbitrary, we have

$$\|H^{11} - U^1 * Q^0 * V^1\|_1 \leq \gamma.$$

Similarly for any $n > 0$ and for any $n_m > n$, $\|P_n(H^{22} - U^2 * Q^{n_m} * V^2)\|_2^2 \le \nu$. Again, this implies that $\|P_n(H^{22} - U^2 * Q^0 * V^2)\|_2^2 \le \nu$. Since n is arbitrary, it follows that

$$\|H^{22} - U^2 * Q^0 * V^2\|_2^2 \le \nu.$$

It follows that Q^0 is an optimal solution for (8.1).

To prove that $\nu_n \nearrow \nu$, we note that $\|P_n(H^{22} - U^2 * Q^{n_m} * V^2)\|_2^2 \le \|P_{n_m}(H^{22} - U^2 * Q^{n_m} * V^2)\|_2^2 = \nu_{n_m}$ for all $n > 0$, for all $n_m > n$.

Taking the limit as m goes to infinity we have

$$\|P_n(H^{22} - U^2 * Q^0 * V^2)\|_2^2 \le \lim_{m\to\infty} \nu_{n_m} \text{ for all } n > 0.$$

It follows that

$$\|H^{22} - U^2 * Q^0 * V^2\|_2^2 \le \lim_{m\to\infty} \nu_{n_m}.$$

Thus we have shown that $\lim_{m\to\infty} \nu_{n_m} = \nu$. Since ν_n is a monotonically increasing sequence, it follows that $\nu_n \nearrow \nu$.

It is clear from Lemma 8.1.1 that $\Phi^{22,o} := H^{22} - U^2 * Q^0 * V^2$ is unique. If $\Phi^{22,n} := P_n(H^{22} - U^2 * Q^n * V^2)$ then from the discussion above it follows that $\nu_{n_m} = \|\Phi^{22,n_m}\|_2^2$ converges to $\nu = \|\Phi^{22,o}\|_2^2$. Also, $\Phi^{22,n_m}(t)$ converges to $\Phi^{22,o}(t)$. It follows from Lemma 8.2.1 that

$$\|\Phi^{22,n_m} - \Phi^{22,o}\|_2 \to 0 \text{ as } m \to \infty.$$

From Lemma 8.1.1 we also have that if $\hat{U}^2$ and $\hat{V}^2$ have full normal column and row ranks respectively then Q^0 is unique. From the uniqueness of Q^0 it follows that the original sequence, $\{\Phi^{22,n}\}$ converges to $\Phi^{22,o}$ in the two norm. This proves the theorem. □

8.2.2 Converging Upper Bounds

Let $\nu^n(\gamma)$ be defined by

$$\begin{aligned} &\inf_{Q\in\ell_1^{n_u\times n_y}} \|H^{22} - U^2 * Q * V^2\|_2^2 \\ &\text{subject to} \\ &\|H^{11} - U^1 * Q * V^1\|_1 \le \gamma \\ &\|Q\|_1 \le \alpha \\ &Q(k) = 0 \text{ if } k > n. \end{aligned} \tag{8.7}$$

The following theorem shows that $\{\nu^n(\gamma)\}$ defines a sequence of upper bounds to $\nu(\gamma)$ which converge to $\nu(\gamma)$.

Theorem 8.2.2. *For all n, $\nu^n(\gamma) \ge \nu^{n+1}(\gamma) \ge \nu(\gamma)$. Also,*

$$\nu^n(\gamma) \searrow \nu(\gamma).$$

Proof. It is clear that $\nu^n(\gamma) \geq \nu^{n+1}(\gamma)$ because any Q in $\ell_1^{n_u \times n_y}$ which satisfies the constraints in the problem definition of $\nu^n(\gamma)$ will satisfy the constraints in the problem definition of $\nu^{n+1}(\gamma)$. For the same reason we also have $\nu^n(\gamma) \geq \nu(\gamma)$ for all relevant n.

Thus $\{\nu^n(\gamma)\}$ is a decreasing sequence of real numbers bounded below by $\nu(\gamma)$. It can be shown that $\nu(\gamma)$ is a continuous function of γ (see Theorem 6.5 in [15]).

Given $\epsilon > 0$ choose $\delta > 0$ such that

$$\nu(\gamma - \delta) - \nu(\gamma) < \frac{\epsilon}{2}. \tag{8.8}$$

Such a δ exists from the continuity of $\nu(\gamma)$ in γ. Let $Q^{\gamma-\delta}$ be a solution to the problem $\nu(\gamma - \delta)$ which is guaranteed to exist from Theorem 8.2.1. Let M be large enough so that $m \geq M$ implies that

$$|\|H^{22} - U^2 * P_m(Q^{\gamma-\delta}) * V^2\|_2^2 - \|H^{22} - U^2 * Q^{\gamma-\delta} * V^2\|_2^2| < \frac{\epsilon}{2} \text{ and} \tag{8.9}$$

$$|\|H^{11} - U^1 * P_m(Q^{\gamma-\delta}) * V^1\|_1 - \|H^{11} - U^1 * Q^{\gamma-\delta} * V^1\|_1| < \frac{\delta}{2}. \tag{8.10}$$

As $Q^{\gamma-\delta}$ is a solution to the problem $\nu(\gamma - \delta)$ it is also true that

$$\begin{aligned} \|H^{22} - U^2 * Q^{\gamma-\delta} * V^2\|_2^2 &= \nu(\gamma - \delta), \\ \|H^{11} - U^1 * Q^{\gamma-\delta} * V^1\|_1 &\leq \gamma - \delta \text{ and} \\ \|Q^{\gamma-\delta}\|_1 &\leq \alpha. \end{aligned}$$

From the above and equations (8.9), (8.10) it follows that for all $m \geq M$,

$$\|H^{22} - U^2 * P_m(Q^{\gamma-\delta}) * V^2\|_2^2 - \nu(\gamma - \delta) \leq \frac{\epsilon}{2}, \tag{8.11}$$

$$\|H^{11} - U^1 * P_m(Q^{\gamma-\delta}) * V^1\|_1 \leq \gamma \text{ and} \tag{8.12}$$

$$\|P_m(Q^{\gamma-\delta})\|_1 \leq \alpha. \tag{8.13}$$

From equation (8.8) and the above it follows that for all $m \geq M$, $P_m(Q^{\gamma-\delta})$ satisfies all the constraints of problem $\nu^m(\gamma)$ and

$$\|H^{22} - U^2 * P_m(Q^{\gamma-\delta}) * V^2\|_2^2 - \frac{\epsilon}{2} - \nu(\gamma) \leq \frac{\epsilon}{2}.$$

Thus for all $m \geq M$ it follows that

$$\nu^m(\gamma) - \nu(\gamma) \leq \epsilon.$$

This proves the theorem. □

8.3 Summary

In this chapter we have formulated a problem which incorporates the $\mathcal{H}_2$ performance measure and the ℓ_1 measure. It is shown that converging upper and lower bounds can be obtained via finite dimensional convex programming problems. This methodology avoids many of the problems present in zero interpolation based methods.

9. Robust Performance

The robust stability problem addresses the stability of the closed loop system for all perturbations Δ which lie in a specified class. The larger this class, the more conservative the condition on the closed loop map which guarantees stability with respect to the perturbations in the class. The small gain theorem gives a condition on the closed loop map which is sufficient for stability when the only restriction on the perturbation is a norm bound. However, many physical systems can be cast into the framework of Figure 9.1 with Δ having a block diagonal structure [3, 25]. This has led to research into conditions on the closed loop map for robust stability with respect to perturbations which have a block diagonal structure. Necessary and sufficient conditions for robust stability with respect to linear time varying and block diagonal perturbations which have finite ℓ_∞ induced norms are given in [26, 27, 28]. Similar conditions can be obtained when the perturbations are nonlinear time invariant instead of linear time varying. These conditions on the closed loop map can be verified easily and can be considerably less conservative than the small gain condition.

The robust performance problem in contrast to the nominal performance problem addresses the issue of synthesis of a controller K which minimizes the effect of w on z over all controllers which stabilize the system in Figure 9.1 for the worst case Δ belonging to a specified class. The ℓ_1 robust performance problem captures the objectives of ℓ_1 robust stability and ℓ_1 performance. In [29] it is shown that such a problem can be solved by employing sensitivity methods to linear programming, when there is only one perturbation block. However, as mentioned earlier ℓ_1 performance is no guarantee of acceptable $\mathcal{H}_2$ performance. Motivated by these concerns we formulate a problem which reflects the objectives of the ℓ_1 robust design and nominal $\mathcal{H}_2$ performance. We show that this problem can be solved via finite dimensional quadratic programming when there is only one perturbation block.

This chapter is organized as follows. In Section 9.1 we give results on robust stability and robust performance when the perturbation block is bounded in the ℓ_∞ induced norm. In Section 9.2 we present the problem of interest and give upper and lower bounds to the main problem. We also show the connection between these bounds and quadratic programming. In Section 9.3 we give results on quadratic programming while in Section 9.4

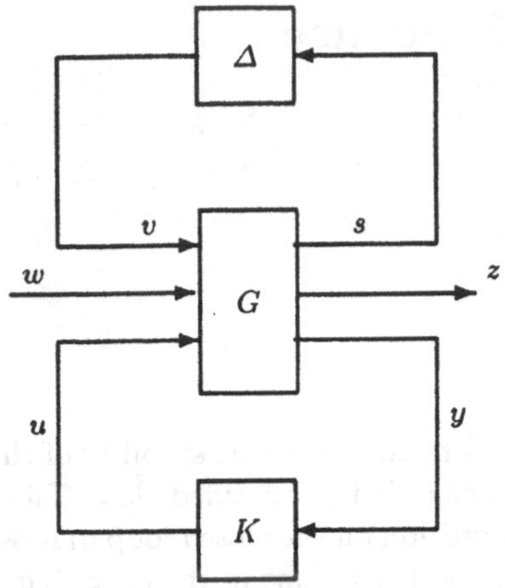

Fig. 9.1. The Performance Problem

we describe the solution method to the problem formulated in this chapter. Finally, in Section 9.5 we give a summary of this chapter.

9.1 Robust Stability and Robust Performance

In the first part of this section we present results on robust stability. In the second part we present results on robust performance.

9.1.1 Robust Stability

Let Δ be a causal map from ℓ_∞ to ℓ_∞ and

$$||\Delta||_{\infty-ind} := \sup_{||w||_\infty \leq 1} ||\Delta w||_\infty.$$

The small gain theorem tells us that the system in Figure 9.2(a) is stable for all $\Delta \in \{\Delta \text{ causal} : \; ||\Delta||_{\infty-ind} \leq 1\}$ if $||M||_1 < 1$. This class of perturbations does not include any structure. However, many physical systems have perturbations which have diagonal structure as shown in Figure 9.2(b). The small gain theorem is a conservative result for such systems. Therefore, it is natural to ask when the system is stable for a restricted class of perturbations which accounts for the structure. Consider Figure 9.2(b), where M is in $\ell_1^{n\times n}$ and Δ_i are single-input single-output maps for $i = 1, \ldots, n$. Define,
$\Delta_{LTV} := \{\Delta |\; \Delta = diag\{\Delta_1, \ldots, \Delta_n\},\; \Delta_i$ is causal linear time varying$\}$,
$\Delta_{NL} := \{\Delta |\; \Delta = diag\{\Delta_1, \ldots, \Delta_n\},\; \Delta_i$ is causal nonlinear time invariant$\}$,

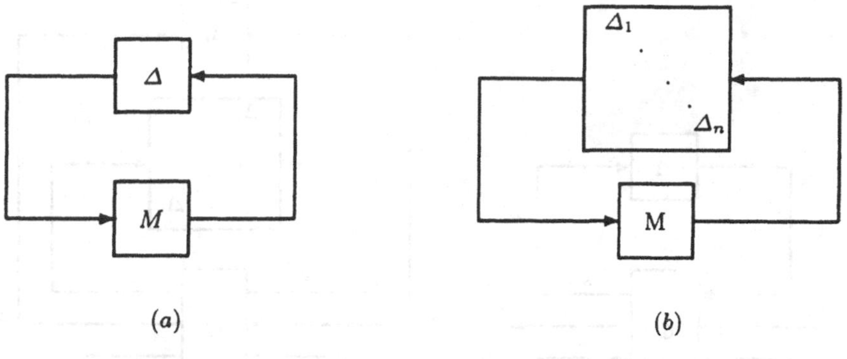

Fig. 9.2. (a) Perturbations are unstructured. (b) Perturbations have diagonal structure

$\mathbf{B}_{\Delta_{LTV}} := \{\Delta \in \Delta_{LTV} | \ ||\Delta||_{\infty-ind} \leq 1\}$ and
$\mathbf{B}_{\Delta_{NL}} := \{\Delta \in \Delta_{NL} | \ ||\Delta||_{\infty-ind} \leq 1\}$.
$\mathbf{B}_{\Delta_{LTV}}$ and $\mathbf{B}_{\Delta_{NL}}$ incorporate the diagonal structure of the perturbations. Each of these sets leads to a question of robust stability. We are interested in a condition on M which if satisfied leads to the stability of the system for all perturbations lying in $\mathbf{B}_{\Delta_{LTV}}$ and a condition on M which leads to stability of the system for all perturbations in $\mathbf{B}_{\Delta_{NL}}$. The Theorem 9.1.1 proven in [27] gives such conditions. We define $|M|$ where M in $\ell_1^{n\times n}$ by the following,

$$|M| := \begin{pmatrix} ||M_{11}||_1 & \dots & ||M_{1n}||_1 \\ \vdots & \vdots & \vdots \\ ||M_{n1}||_1 & \dots & ||M_{nn}||_1 \end{pmatrix}.$$

For the rest of this chapter by stability we mean ℓ_∞ stability.

Theorem 9.1.1. *The system in Figure 9.2(b) is stable for all $\Delta \in \mathbf{B}_{\Delta_{LTV}}$ if and only if either one of the following conditions is satisfied*

1) $\rho(|M|) < 1$ where $\rho(.)$ is the spectral radius,
2) $\inf_{D\in\mathcal{D}} ||D^{-1}MD||_1 < 1$

where $\mathcal{D}$ is the set of diagonal matrices with strictly positive elements. The same result holds if Δ is restricted to lie in $\mathbf{B}_{\Delta_{NL}}$ instead of $\mathbf{B}_{\Delta_{LTV}}$.

9.1.2 Robust Performance

Consider Figure 9.3(a). The ℓ_1 robust performance problem with respect to linear time varying structured perturbations is the problem of synthesizing a controller K such that the closed loop map is stable for all $\Delta \in \mathbf{B}_{\Delta_{LTV}}$ and the ℓ_1 norm of the map from w to z is less than one for all $\Delta \in \mathbf{B}_{\Delta_{LTV}}$. In [27] it is shown that this problem is the same as the problem of synthesizing a

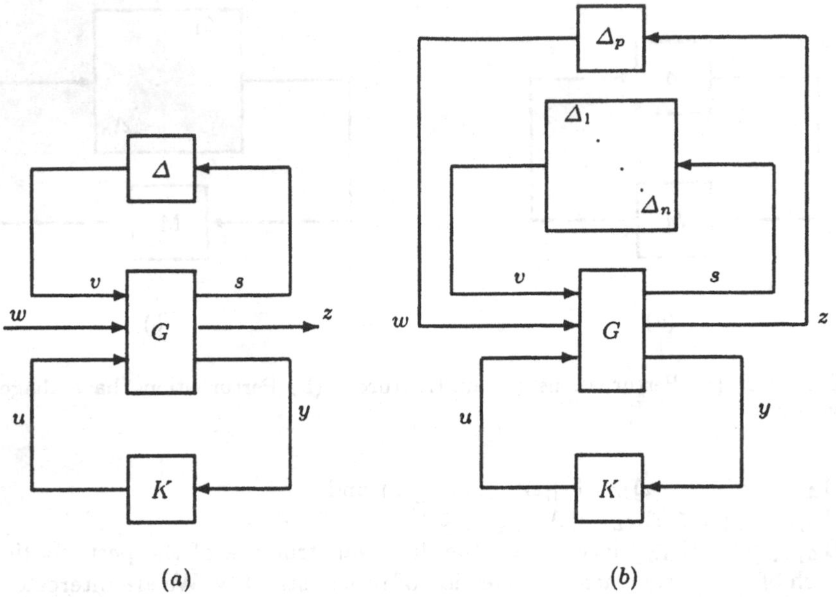

Fig. 9.3. Robust Performance for (*a*) is equivalent to robust stability of (*b*)

controller K such that the system in Figure 9.3(*b*) is stable for every (Δ, Δ_p) which lies in $\mathbf{B}_{\Delta_p}$, which is the set $\{\overline{\Delta} = diag(\Delta,\ \Delta_p) |\ \Delta \in \mathbf{B}_{\Delta_{LTV}},\ ||\Delta_p||_{\infty-ind} \leq 1,\ \ \Delta_p \text{ is } LTV \text{ and causal}\}$. This results in the following. theorem [28],

Theorem 9.1.2. *The system in Figure 9.3 achieves robust performance with respect to* $\mathbf{B}_{\Delta_{LTV}}$ *if and only if either of the following conditions is satisfied*

$$(a)\ \ \rho(|\Phi|) < 1,$$
$$(b)\ \ \inf_{D \in \mathcal{D}} ||D^{-1}\Phi D||_1 < 1,$$

where Φ *is the closed loop map from* $\begin{pmatrix} v \\ w \end{pmatrix}$ *to* $\begin{pmatrix} s \\ z \end{pmatrix}$ *and* $\mathcal{D}$ *is the set of diagonal matrices with strictly positive elements which have compatible dimensions with the dimensions of* Φ.

The same theorem holds for the robust performance problem with respect to nonlinear time invariant structured perturbations. The extra block Δ_p is now restricted to be nonlinear time invariant.

9.2 Problem Formulation

In this section we define the optimization problem which captures the objectives of interest stated in the previous sections. After defining the problem of interest we formulate problems which give upper and lower bounds to the problem of interest.

Youla parametrization (see Theorem 4.2.4) tells us that all closed loop maps achievable through stabilizing controllers are given by

$$\Phi = H - U * Q * V,$$

where H, U and V are fixed elements dependent only on the system G and Q is a stable free parameter. In this chapter we assume the closed loop map is a two-input two-output map. We define the sets $\mathcal{D}$ and $\Theta(D)$ to be the sets

$\mathcal{D} := \{\begin{pmatrix} d_1 & 0 \\ 0 & d_2 \end{pmatrix} : d_i > 0\}$ and ,

$\Theta(D) := \{\Phi \in \ell_1^{2\times 2} | \Phi = H - U * Q * V$ for Q stable $||D^{-1}\Phi D||_1 \leq \gamma\}$,

respectively. Let

$$\mu(D) := \inf\{||\Phi_{22}||_2{}^2 \mid \Phi \in \Theta(D)\}, \tag{9.1}$$

$$\mu := \inf_{D \in \mathcal{D}} \mu(D). \tag{9.2}$$

We Note that Φ_{22} is the map between w and z. μ captures the objectives we have in mind. In the following sections we obtain converging upper and lower bounds to μ.

9.2.1 Delay Augmentation Approach

Let n_u, n_w, n_v, n_y, n_z and n_s denote the dimension of u, w, v, y, z and s respectively. The Youla parametrization tells us that all achievable closed loop maps are given by $H - U * Q * V$, where H, U and V are fixed stable elements and Q is a free variable which is stable (for a detailed discussion see [3]). The constraint that Q is stable can be translated into linear constraints on Φ. Thus there exists an operator $\mathcal{A}$: $\ell_1^{n_w+n_v} \to \ell_1$ such that $\Phi = H - U * Q * V$ for some Q stable if and only if $\mathcal{A}(\Phi) = b$, where b is a fixed element in ℓ_1. The range space of the operator $\mathcal{A}$ is finite dimensional if $n_z + n_s = n_u$ and $n_w + n_v = n_y$. In this case the problem is called a *square problem*. We define

$$\mu^\delta(D) := \inf_{\Phi \in \Theta(D)} \{||\Phi_{22}||_2{}^2 + \delta(||(D^{-1}\Phi D)_1||_1 + ||(D^{-1}\Phi D)_2||_1)\}, \tag{9.3}$$

$$\mu^\delta := \inf_{D \in \mathcal{D}} \mu^\delta(D), \tag{9.4}$$

where $(D^{-1}\Phi D)_i$ is the ith row of Φ. It is clear that we can approximate μ by μ^δ to the desired accuracy by choosing an appropriate δ. $\mu(D)$ is a quadratic programming problem. However, it is intractable because it is not possible to show that it can be solved via *finite* dimensional quadratic programming even in the square case. In contrast we have seen in Chapter 7 that for the square case, $\mu^\delta(D)$ can be solved via finite dimensional quadratic programming.

If $n_z + n_s > n_u$ or $n_w + n_v > n_y$ then the range space of $\mathcal{A}$ is not finite dimensional. In this case $\mu^\delta(D)$ is solved by converting it to a square problem.

This is done by the *Delay Augmentation Method.* We give a brief description of this method here (for a detailed discussion see [3]). Let S denote a unit shift, that is,

$$S(x(0), x(1), x(2), \ldots) = (0, x(0), x(1), \ldots),$$

and S^T denotes a Tth order shift. Suppose, that the Youla parametrization of the plant yields H in $\ell_1^{n_z' \times n_w'}$, U in $\ell_1^{n_z' \times n_u}$, and V in $\ell_1^{n_y \times n_w'}$, where $n_z' = n_z + n_s$ and $n_w' = n_w + n_v$. Partition, $\hat{U}$ into

$$\hat{U} = \begin{pmatrix} \hat{U}^1 \\ \hat{U}^2 \end{pmatrix},$$

where U^1 in $\ell_1^{n_u \times n_u}$. Similarly, partition $\hat{V}$ into $(\hat{V}^1, \hat{V}^2)$ where V^1 in $\ell_1^{n_y \times n_y}$. Let Φ and H be partitioned according to the following equation:

$$\begin{pmatrix} \hat{\Phi}^{11} & \hat{\Phi}^{12} \\ \hat{\Phi}^{21} & \hat{\Phi}^{22} \end{pmatrix} = \begin{pmatrix} \hat{H}^{11} & \hat{H}^{12} \\ \hat{H}^{21} & \hat{H}^{22} \end{pmatrix} - \begin{pmatrix} \hat{U}^1 \\ \hat{U}^2 \end{pmatrix} \hat{Q}^{11} \begin{pmatrix} \hat{V}^1 & \hat{V}^2 \end{pmatrix}. \tag{9.5}$$

We augment $\hat{U}$ and $\hat{V}$ by N^{th} order delays and augment $\hat{Q}^{11}$ as given by the following

$$\begin{pmatrix} \hat{\Phi}^{11,N} & \hat{\Phi}^{12,N} \\ \hat{\Phi}^{21,N} & \hat{\Phi}^{22,N} \end{pmatrix} = \begin{pmatrix} \hat{H}^{11} & \hat{H}^{12} \\ \hat{H}^{21} & \hat{H}^{22} \end{pmatrix} - \begin{pmatrix} \hat{U}^1 & 0 \\ \hat{U}^2 & \hat{S}^N \end{pmatrix} \begin{pmatrix} \hat{Q}^{11} & \hat{Q}^{12} \\ \hat{Q}^{21} & \hat{Q}^{22} \end{pmatrix} \begin{pmatrix} \hat{V}^1 & \hat{V}^2 \\ 0 & \hat{S}^N \end{pmatrix} \tag{9.6}$$

or equivalently,

$$\hat{\Phi}^N := \hat{H} - \hat{U}^N \hat{Q} \hat{V}^N.$$

We define $\Theta(D, N)$ the feasible set for the delay augmented problem to be the set

$$\{\Phi^N \in \ell_1^{2\times 2} |\ \hat{\Phi}^N = \hat{H} - \hat{U}^N \hat{Q} \hat{V}^N \text{ with } \hat{Q} \text{ stable and } \|D^{-1}\Phi^N D\|_1 \le \gamma\}.$$

We define the Delay Augmentation problem of order N by

$$\mu_N^\delta(D) := \inf\{l(\Phi^N) : \Phi^N \in \Theta(D, N)\}. \tag{9.7}$$

where

$$l(\Phi) := \|\Phi_{22}\|_2{}^2 + \delta(\|(D^{-1}\Phi D)_1\|_1 + \|(D^{-1}\Phi D)_2\|_1).$$

This is a square problem and can be solved via finite dimensional quadratic programming. Let the delay augmented problem corresponding to (9.4) be given by

$$\mu_N^\delta := \inf_{D \in \mathcal{D}} \mu_N^\delta(D). \tag{9.8}$$

We will now show that μ_N^δ converges to μ^δ from below.

Lemma 9.2.1.

$$\lim_{N\to\infty} \mu_N^\delta = \mu^\delta,$$

where the limit on the left hand side of the equation above exists.

Proof. It can be shown that $\Theta(D) \subset \Theta(D, N+1) \subset \Theta(D, N)$ for all integers N. Therefore, for a given D and for all N

$$\mu_N^\delta(D) \leq \mu_{N+1}^\delta(D) \text{ and } \mu_N^\delta(D) \leq \mu^\delta(D).$$

Therefore $\mu_N^\delta(D)$ is an increasing sequence in N. It can be shown that this sequence converges to $\mu^\delta(D)$ from below. Now, $\mu_N^\delta(D) \leq \mu^\delta(D)$ for all D in $\mathcal{D}$. This implies that

$$\inf_{D\in\mathcal{D}} \mu_N^\delta(D) \leq \inf_{D\in\mathcal{D}} \mu^\delta(D).$$

Therefore, $\mu_N^\delta \leq \mu^\delta$. Also, because $\mu_N^\delta(D) \leq \mu_{N+1}^\delta(D)$ for all D in $\mathcal{D}$ it follows that μ_N^δ is an increasing sequence bounded above by μ^δ. Let $L = \lim_{N\to\infty} \mu_N^\delta$. Suppose $\mu^\delta - L =: \epsilon > 0$. Then, there exists an integer M such that

$$\mu^\delta - \mu_N^\delta \geq \epsilon/2 \ \forall\ N \geq M.$$

This implies that

$$\mu^\delta - (\mu_N^\delta + \epsilon/4) \geq \epsilon/4 \ \forall\ N \geq M.$$

Therefore there exists D_0 in $\mathcal{D}$ such that

$$\mu^\delta(D_0) - \mu_N^\delta(D_0) \geq \epsilon/4 \ \forall\ N \geq M.$$

This, contradicts the fact that $\mu_N^\delta(D_0) \to \mu^\delta(D_0)$ as $N \to \infty$. This proves the lemma. □

Notice that for a given D in $\mathcal{D}$ we have the following:

$$D^{-1}\Phi D = \begin{pmatrix} \Phi_{11} & (d_2/d_1)\Phi_{12} \\ (d_1/d_2)\Phi_{21} & \Phi_{22} \end{pmatrix}.$$

We denote d_2/d_1 by s and therefore we have

$$D^{-1}\Phi D = \begin{pmatrix} \Phi_{11} & s\Phi_{12} \\ (1/s)\Phi_{21} & \Phi_{22} \end{pmatrix}.$$

Thus, $\mu_N^\delta(D)$ can be obtained by solving the following problem.

$$\inf_{\Phi \text{ Achievable}} ||\Phi_{22}||_2{}^2 + \delta(||\Phi_{11}||_1 + s||\Phi_{12}||_1 + 1/s||\Phi_{21}||_1 + ||\Phi_{22}||_1)$$

subject to

$$||\Phi_{11}||_1 + s||\Phi_{12}||_1 \leq \gamma$$
$$1/s||\Phi_{21}||_1 + ||\Phi_{22}||_1 \leq \gamma.$$

where Φ is achievable if $\hat{\Phi} := \hat{H} - \hat{U}^N\hat{Q}\hat{V}^N$ for some $\hat{Q}$ stable. It can be shown that the above problem is a finite dimensional quadratic programming problem (since it is a square problem) where the dimension can be determined *a priori* (see Chapter 7). We assume in this section that the dimension is given by T. We define new variables corresponding to each $\Phi_{ij}(t)$ as follows:

$$\begin{aligned}\Phi^+_{ij}(t) &:= \Phi_{ij}(t) \text{ if } \Phi_{ij}(t) \geq 0\\ &:= 0 \text{ if } \Phi_{ij}(t) < 0\\ \Phi^-_{ij}(t) &:= -\Phi_{ij}(t) \text{ if } \Phi_{ij}(t) \leq 0\\ &:= 0 \text{ if } \Phi_{ij}(t) > 0.\end{aligned}$$

This implies that $\Phi_{ij}(t) = \Phi^+_{ij}(t) - \Phi^-_{ij}(t)$ and $|\Phi_{ij}(t)| = \Phi^+_{ij}(t) + \Phi^-_{ij}(t)$. Also, we can obtain $\mu^\delta_N(D)$ by solving the following.

$$\begin{aligned}\min_{\Phi \text{ Achievable}} \sum_{t=0}^{T} &(\Phi^+_{22}(t) - \Phi^-_{22}(t))^2 + \delta(\Phi^+_{11}(t) + \Phi^-_{11}(t) + \Phi^+_{22}(t) + \Phi^-_{22}(t))\\ &+\delta(s\Phi^+_{12}(t) + s\Phi^-_{12}(t) + 1/s\Phi^+_{21}(t) + 1/s\Phi^-_{21}(t))\end{aligned}$$

subject to

$$\sum_{t=0}^{T} \Phi^+_{11}(t) + \Phi^-_{11}(t) + s\Phi^+_{12}(t) + s\Phi^-_{12}(t) \leq \gamma,$$

$$\sum_{t=0}^{T} \Phi^+_{22}(t) + \Phi^-_{22}(t) + 1/s\Phi^+_{21}(t) + 1/s\Phi^-_{21}(t) \leq \gamma,$$

$$\Phi^+_{ij}(t) \geq 0,\ \Phi^-_{ij}(t) \geq 0,\quad \forall\, t = 0, \ldots, T.$$

Let

$$E := \begin{pmatrix} 0_{6(T+1)+2} \\ I_{T+1} \\ -I_{T+1} \end{pmatrix}$$

and let $C = 2EE'$. We define $x_1 := \sum_{t=0}^{T} \Phi^+_{12}(t) + \Phi^-_{12}(t)$ and $x_2 := \sum_{t=0}^{T} \Phi^+_{21}(t) + \Phi^-_{21}(t)$. Let the vector x and $p(s)$ be given by:

$$x := \begin{pmatrix} x_1 \\ x_2 \\ \Phi^+_{11} \\ \Phi^-_{11} \\ \Phi^+_{12} \\ \Phi^-_{12} \\ \Phi^+_{21} \\ \Phi^-_{21} \\ \Phi^+_{22} \\ \Phi^-_{22} \end{pmatrix} \qquad p(s) := -\delta \begin{pmatrix} s \\ 1/s \\ \mathbf{1} \\ \mathbf{1} \\ \mathbf{0} \\ \mathbf{0} \\ \mathbf{0} \\ \mathbf{0} \\ \mathbf{1} \\ \mathbf{1} \end{pmatrix},$$

where $\mathbf{1}$ is a vector of ones with length $(T+1)$ and $\mathbf{0}$ is a vector of zeros with length $(T+1)$. With the above definitions we can cast $\mu^\delta_N(D)$ into the following form:

$$\begin{aligned}
&\min \frac{1}{2}x'Cx - p'(s)x \\
&\text{subject to} \\
&\begin{pmatrix} A_{11}(s) & A_{12} \\ A_{21} & A_{22} \end{pmatrix} x \leq b \\
&Hx = e \\
&x \geq 0
\end{aligned} \qquad (QP1)$$

where $A_{11}(s) = \begin{pmatrix} s & 0 \\ 0 & \frac{1}{s} \end{pmatrix}$ and $s > 0$. The achievability conditions are absorbed into H. C is positive semidefinite. $\mu_N^\delta(D)$ is a function of the variable s. Therefore, we denote $\mu_N^\delta(D)$ by $\gamma(s)$. Note that

$$\mu_N^\delta = \inf_{s \in R^+} \gamma(s) =: \gamma_{opt}.$$

9.2.2 Finitely Many Variables Approach

Converging upper bounds can be obtained by Finitely Many Variables approach. In this approach the allowable closed loop maps achievable via stabilizing controllers are restricted to have a finite impluse response structure. This means that the allowable closed loop maps can be characterized by the linear map $\mathcal{A}$ as defined in the earlier subsection with $\mathcal{A}$ having a finite dimensional domain space. It can be shown that in this case the range space is also finite dimensional [3]. Thus the allowable closed loop maps can be characterized by finite number of constraints involving finite number of variables. We define

$$\mu^N(D) = \inf\{||\Phi_{22}||_2{}^2 \mid \Phi \text{ in } \Theta^N(D)\}, \qquad (9.9)$$

where $\Theta^N(D) = \{\Phi = H - U * Q * V$ with Q stable$\mid \Phi(N+k) = 0$ for all k and $||D^{-1}\Phi D||_1 \leq \gamma\}$. It can be shown that $\mu^N := \inf_{D \in \mathcal{D}} \mu^N(D) \rightarrow \mu$ from above as $N \rightarrow \infty$ following similar arguments employed in the previous subsection. Defining x and C as defined in the earlier subsection we can cast $\mu^T(D)$ into the following form

$$\begin{aligned}
&\min \frac{1}{2}x'Cx \\
&\text{subject to} \\
&\begin{pmatrix} A_{11}(s) & A_{12} \\ A_{21} & A_{22} \end{pmatrix} x \leq b \\
&Hx = e \\
&x \geq 0
\end{aligned} \qquad (QP2)$$

Note that if we denote $\mu^T(D)$ by $\gamma(s)$ then

$$\mu^T = \inf_{s \in R^+} \gamma(s) =: \gamma_{opt}.$$

9.3 Quadratic Programming

Consider the following quadratic programming problem

$$\begin{aligned} &\min \frac{1}{2}x'Cx - p'x \\ &\text{subject to} \\ &\quad Ax \leq b \\ &\quad Hx = e \\ &\quad x \geq 0 \end{aligned} \qquad (QP)$$

where A in $R^{m_1 \times n_1}$, H in $R^{m_2 \times n_1}$ has full row rank, and C is positive semi-definite. We are interested in obtaining necessary and sufficient conditions for x_0 to be optimal for (QP). The following theorem gives such conditions.

Theorem 9.3.1. *Consider the quadratic programming problem, (QP). x_0 is optimal for the problem if and only if there exist y_0 in R^{m_1}, u in R^{m_1}, λ in R^{m_2}, v in R^{n_1} such that x_0, y_0, u, v, λ satisfy the following conditions*

$$\begin{aligned} p &= Cx_0 + A'u + H'\lambda - v \\ e &= Hx_0 \\ b &= Ax_0 + y_0 \\ 0 &= u'y_0 \\ 0 &= v'x_0 \\ x_0 &\geq 0, y_0 \geq 0, u \geq 0, v \geq 0. \end{aligned}$$

Proof. ($\Rightarrow$) Suppose, x_0 is optimal for the problem (QP). This implies that x_0 satisfies the conditions $e = Hx_0, Ax_0 - b \leq 0$ and $x_0 \geq 0$. From Theorem 3.3.2 (Kuhn-Tucker-Lagrange duality result) we know that there exists u in R^{m_1}, λ in R^{m_2}, v in R^{n_1} with $u \geq 0$ and $v \geq 0$ such that x_0 minimizes $L(x)$ where

$$L(x) := \frac{1}{2}x'Cx - p'x + u'(Ax - b) + \lambda'(Hx - e) - v'x.$$

This implies that

$$\left.\frac{d}{dx}L(x)\right|_{x=x_0} = Cx_0 - p + A'u + H'\lambda - v = 0.$$

Also, from Theorem 3.3.2 we know that

$$u'(Ax_0 - b) + \lambda'(Hx_0 - e) - v'x_0 = 0.$$

As x_0 satisfies $e - Hx_0 = 0$ we have $u'(Ax_0 - b) - v'x_0 = 0$. However, x_0 satisfies $Ax_0 - b \leq 0$ and $x_0 \geq 0$. Therefore we conclude that $u'(Ax_0 - b) = v'x_0 = 0$. The necessity of the conditions given in the theorem statement for x_0 to be optimal is established by defining $y_0 = b - Ax_0$.

($\Leftarrow$) Suppose, for a given x_0 there exist vectors λ, y_0, u, v which satisfy the conditions given in the theorem statement. Let x in R^{n_1} be any element which satisfies the constraints of (QP). Let $f(.)$ denote the objective function of (QP). We have

$$\begin{aligned}
f(x) - f(x_0) &= \frac{1}{2}x'Cx - \frac{1}{2}x_0'Cx_0 - p'(x - x_0) \\
&= \frac{1}{2}(x - x_0)'C(x - x_0) + x'Cx_0 - x_0'Cx_0 - p'(x - x_0) \\
&= \frac{1}{2}(x - x_0)'C(x - x_0) + (x - x_0)'(p - A'u - H\lambda + v) \\
&\qquad -p'(x - x_0) \\
&\geq -(x - x_0)'A'u - (x - x_0)'H'\lambda + (x - x_0)'v \\
&= -u'((Ax - b) - (Ax_0 - b)) - \lambda'H(x - x_0) + (x - x_0)'v \\
&= -u'((Ax - b) + y_0) - \lambda'H(x - x_0) + (x - x_0)'v \\
&= u'(b - Ax) + v'x \geq 0
\end{aligned}$$

This proves the theorem. □

The above theorem shows that the solution of a convex quadratic programming problem as given in (QP) is equivalent to the search of a vector (x, u, v, y, λ) which satisfies the following conditions:

$$\begin{pmatrix} C & A' & -I & 0 & H' \\ A & 0 & 0 & I & 0 \\ H & 0 & 0 & 0 & 0 \end{pmatrix} \begin{pmatrix} x \\ u \\ v \\ y \\ \lambda \end{pmatrix} = \begin{pmatrix} p \\ b \\ e \end{pmatrix}, \tag{9.10}$$

$$v'x + u'y = 0, \tag{9.11}$$

$$(x\ u\ v\ y) \geq 0. \tag{9.12}$$

Also, note that if conditions (9.10) and (9.11) are satisfied then the objective function $f(.)$ of (QP) is given by:

$$\begin{aligned}
f(x) &= \frac{1}{2}x'Cx - p'x \\
&= \frac{1}{2}x'(p - A'u - H'\lambda + v) - p'x \\
&= -\frac{1}{2}p'x - \frac{1}{2}u'Ax - \frac{1}{2}\lambda'Hx + \frac{1}{2}x'v \qquad (9.13) \\
&= -\frac{1}{2}p'x - \frac{1}{2}u'(b - y) - \frac{1}{2}\lambda'e + \frac{1}{2}v'x \\
&= -\frac{1}{2}p'x - \frac{1}{2}b'u - \frac{1}{2}e'\lambda + \frac{1}{2}v'x + \frac{1}{2}u'y \\
&= -\frac{1}{2}p'x - \frac{1}{2}b'u - \frac{1}{2}e'\lambda \qquad (9.14)
\end{aligned}$$

Define $\mathbf{b} := \begin{pmatrix} p \\ b \\ e \end{pmatrix}$ and $\mathbf{x} := \begin{pmatrix} x \\ u \\ v \\ y \\ \lambda \end{pmatrix}$. Let the matrix in equation (9.10) be denoted by $\mathbf{A}$. Also, we assume that $\mathbf{A}$ in $R^{m\times n}$, has full row rank (i.e. it has rank m). Note that the objective function $f(.)$ of (QP) is given by

$$\begin{aligned} f &= (-\tfrac{1}{2}p \; -\tfrac{1}{2}b \; 0 \; 0 \; -\tfrac{1}{2}e)'\mathbf{x} \\ &=: c'\mathbf{x} \end{aligned},$$

(see equation (9.14)). In this section, whenever, we refer to $\mathbf{x}$ we assume that it is in the form $(x \;\; u \;\; v \;\; y \;\; \lambda)'$ where the variables x, u, v, y and λ are as defined in Theorem 9.3.1. We call x_i and y_i *primal variables.* We call v_i the *dual variable* associated with the primal variable x_i and u_i as the *dual variable* associated with the primal variable y_i. Before we characterize the set of elememts which satisfy equations (9.10), (9.11) and (9.12), we give the following definitions.

Definition 9.3.1 (Feasible solution). *An element $\mathbf{x}$ in R^n is called feasible if it satisfies equations (9.10), (9.11) and (9.12). The set of all such elements is denoted by $\mathcal{F}$.*

Note that a primal variable and its corresponding dual variable both cannot be nonzero in a feasible solution, because of (9.11) and (9.12).

Definition 9.3.2 (Basic solution). *Let B be a $m \times m$ submatrix formed from the columns of $\mathbf{A}$ such that B is invertible. Then, $\mathbf{x}_B := B^{-1}\mathbf{b}$ defines a basic solution of $\mathbf{Ax} = \mathbf{b}$. Such a solution will have $n - m$ components equal to zero corresponding to the columns of $\mathbf{A}$ not in B. These components are called the non-basic variables. The m components that correspond to the columns of B are called basic variables.*

Definition 9.3.3 (Basic feasible solution). *An element $\mathbf{x}$ in R^n is called basic feasible solution if it is basic and feasible.*

Theorem 9.3.2. *If $\mathcal{F}$ is not empty then it has at least one basic feasible solution.*

Proof. Let a_i denote the ith column of $\mathbf{A}$. Let z be a feasible solution and let the ith element of the vector z be denoted by z_i. Also, let z be partitioned as $(x^z \; u^z \; v^z \; y^z \; \lambda^z)'$ where the variables $x^z, u^z, v^z, y^z, \lambda^z$ correpond to variables x, u, v, y, λ in Theorem 9.3.1 indexed by z. For simplicity assume that the first p components of z are nonzero while the rest are zero. This means that

$$z_1 a_1 + z_2 a_2 + \ldots + z_p a_p = \mathbf{b},$$

and z is such that $(u^z)'y^z + (v^z)'x^z = 0$ and $(x^z \; u^z \; v^z \; y^z)' \geq 0$.

If $a_1, \ldots, a_p$ are independent columns then $p \leq m$ because $\mathbf{A}$ has rank m. This implies that z is a basic solution. Suppose, $a_1, \ldots, a_p$ are dependent. Then there exists α in R^n with at least one strictly positive element such that

$$\alpha_1 a_1 + \alpha_2 a_2 + \ldots + \alpha_p a_p = 0,$$

with the last $n-p$ components equal to zero. Let

$$\epsilon := \min_{\{i:\alpha_i > 0\}} \frac{z_i}{\alpha_i}.$$

Let $t := (z - \epsilon\alpha)$. This implies that $\mathbf{A}t = \mathbf{A}(z - \epsilon\alpha) = \mathbf{b}$ because $\mathbf{A}\alpha = 0$. Also, note that if $z_i = 0$ then $t_i = 0$. The condition $(u^z)'y^z + (v^z)'x^z = 0$ is equivalent to $u_i^z y_i^z = v_i^z x_i^z = 0$ (because $(x^z\ u^z\ v^z\ y^z)' \geq 0$). This means that $u_i^t y_i^t = v_i^t x_i^t = 0$. Also, if $z_i > 0$ then $t_i \geq 0$. Thus t is a feasible solution. From the definition of ϵ, t will have atmost $p-1$ nonzero components. Thus from a feasible solution which had p nonzero components we have created a feasible solution which has $p-1$ nonzero components. This process can be repeated until the number of strictly positive components is less than or equal to m and the corresponding columns of $\mathbf{A}$ are linearly independent (i.e. until the feasible solution is also basic). This concludes the proof. □

In the next section we exploit Theorem 9.3.2 to solve the robust performance problem we have formulated.

9.4 Problem Solution

We saw in Section 9.2 that converging upper and lower bounds to μ as defined in (9.2) can be obtained by solving problems which can be cast into the following form:

$$\begin{aligned} &\min \frac{1}{2}x'Cx - p'(s)x \\ &\text{subject to} \\ &\begin{pmatrix} A_{11}(s) & A_{12} \\ A_{21} & A_{22} \end{pmatrix} x \leq b \\ &Hx = e \\ &x \geq 0 \end{aligned} \qquad (QP(s))$$

with

$$\gamma_{opt} := \inf_{s \in R+} \gamma(s),$$

where γ_{opt} is the problem of interest. Note that $A_{11}(s)$ has the structure given by

$$A_{11}(s) = \begin{pmatrix} s & 0 \\ 0 & \frac{1}{s} \end{pmatrix}$$

and $s > 0$. Using the results obtained in the previous section we know that the above problem has a solution if and only if there exists x, u, v, y, λ which satisfy the following constraints:

$$\begin{pmatrix} C & A'(s) & -I & 0 & H' \\ A(s) & 0 & 0 & I & 0 \\ H & 0 & 0 & 0 & 0 \end{pmatrix} \begin{pmatrix} x \\ u \\ v \\ y \\ \lambda \end{pmatrix} = \begin{pmatrix} p \\ b \\ e \end{pmatrix}, \tag{9.15}$$

$$v'x + u'y = 0, \tag{9.16}$$

$$(x\ u\ v\ y) \geq 0. \tag{9.17}$$

Using the structure of $A(s)$ the constraints given by equation (9.15) can be rearranged as given below:

$$\begin{pmatrix} s & 0 & 0 & 0 & * & * & * \\ 0 & \frac{1}{s} & 0 & 0 & * & * & * \\ * & * & -s & 0 & * & * & * \\ * & * & 0 & -\frac{1}{s} & * & * & * \\ * & * & * & * & * & * & * \\ * & * & * & * & * & * & * \end{pmatrix} \begin{pmatrix} x_1 \\ x_2 \\ u_1 \\ u_2 \\ \underline{x} \\ \lambda \end{pmatrix} = \begin{pmatrix} b_1 \\ b_2 \\ p_1(s) \\ p_2(s) \\ \underline{b} \\ \underline{e} \end{pmatrix}$$

where the entries denoted by $*$ do not depend on s. We denote the matrix on the left hand side by $\mathbf{A}(s)$, the vector on the rightmost side of the equation by $\mathbf{b}(s)$ and $(x_1, x_2, u_1, u_2, \underline{x}, \lambda)'$ by $\mathbf{x}$ (note that λ is the last element in $\mathbf{x}$). We have also shown that if $(QP(s))$ has a finite value for some fixed s then there exists a basic feasible solution z of $\mathbf{A}(s)\mathbf{x} = \mathbf{b}(s)$. Note that the lower bound given by $(QP1)$ and the upper bound given by $(QP2)$ (which are of the form $(QP(s))$ always have a finite value. Thus we will assume that $(QP(s))$ has a finite value for all relevant s. Also note that $f(.)$ is given by

$$f(\mathbf{x}) = c'(s)\mathbf{x}.$$

Suppose, for some fixed value $s_0 > 0$ we have obtained a basic feasible solution of $(QP(s_0))$, given by z_{s_0}. Note that because of condition (9.11) one can choose the matrix $B(s_0)$ in $R^{m \times m}$ where $B(s_0)$ is the associated matrix with the basic solution z_{s_0} (see Definition 9.3.2) such that if a column corresponding to a dual (primal) variable is included in $B(s_0)$ then the column associated with the primal (dual) variable is not in $B(s_0)$. We call the m independent columns of $B(s_0)$ as the *optimal basis* associated with z_{s_0}. Our intention is to characterize the set of reals $0 < s$ such that $(QP(s))$ has a basic feasible solution which has the same optimal basis as the optimal basis of z_{s_0}. The way we have chosen the optimal basis for s_0 guarantees that the condition (9.11) is satisfied if we generate a basic solution using the same columns for a value of s different from s_0 (because the product $v_i x_i = u_i y_i = 0$ will always be true if a solution is generated with the an optimal basis). We introduce some notation now. We assume that $\mathbf{A}(s)$ is a $m \times n$ matrix with $m \leq n$.

Given an indexing set of m positive integers $\mathcal{J} = \{j_1, \ldots, j_m\}$, the notation $B_{\mathcal{J}}(s)$ denotes the matrix formed by those columns of $\mathbf{A}(s)$ indexed by the elements of $\mathcal{J}$. An indexing set is said to be *basis-index* if the $m \times m$ matrix $B_{\mathcal{J}}(s)$ is invertible and is an *optimal-basis-index* if $B_{\mathcal{J}}(s)$ is an optimal basis for the problem $(QP(s))$. The vector c_B in $R^{1\times m}$ consists of entries of c corresponding to the basic variables whereas c_D is the $1 \times (n-m)$ vector corresponding to the nonbasic variables. Let β be defined as $\beta := B_{\mathcal{J}}^{-1} = [\beta_{ij}] = [\beta^1\ \beta^2\ \ldots\ \beta^m]$.

Definition 9.4.1. *Let $s_0 > 0$. Let $\mathcal{J}_0$ be an optimal basis index for the problem $(QP(s_0))$. Define $\mathbf{x}_{\mathcal{J}_0}(.) : R \rightarrow R^m$, the solution function w.r.t $\mathcal{J}$ as follows*

$$\mathbf{x}_{\mathcal{J}_0}(s) := B_{\mathcal{J}_0}^{-1}(s)\mathbf{b}(s)$$

if $B_{\mathcal{J}_0}^{-1}(s)$ exists. Otherwise this function is given a value 0.

We assume throughout this chapter that x_1 and x_2 are basic variables.

Theorem 9.4.1. *Let $s_0 > 0$. Let $\mathcal{J}_0$ be an optimal-basis-index with x_B as the basic feasible solution for the problem $(QP(s_0))$. Suppose u_1 and u_2 are basic variables in the optimal solution. Define $B := B_{\mathcal{J}_0}(s_0)$ and let $\beta := B^{-1}$. Then $B_{\mathcal{J}_0}(s)$ is invertible if and only*

$$\alpha(s) := det(I_4 + SYB^{-1}X) \neq 0$$

where

$$X = \begin{pmatrix} I_4 \\ 0 \end{pmatrix}, \quad S = \begin{pmatrix} s - s_0 & 0 & 0 & 0 \\ 0 & \frac{s_0 - s}{s_0 s} & 0 & 0 \\ 0 & 0 & s_0 - s & 0 \\ 0 & 0 & 0 & \frac{s - s_0}{s_0 s} \end{pmatrix}, \quad Y = (I_4 \quad 0).$$

If $\alpha(s) \neq 0$ then

$$\mathbf{x}_{\mathcal{J}_0}(s) = x_B(s) - \left[\beta^1\ \beta^2\ \beta^3\ \beta^4\right] R(s) \begin{bmatrix} (x_B(s))_1 \\ (x_B(s))_2 \\ (x_B(s))_3 \\ (x_B(s))_4 \end{bmatrix}$$

where

$$R(s) := (I_4 + SYB^{-1}X)^{-1}S \text{ and } x_B(s) := B^{-1}\mathbf{b}(s).$$

If $\alpha(s) = 0$ then $\mathbf{x}_{\mathcal{J}_0}(s) = 0$.

Proof. Let $B := B_{\mathcal{J}_0}(s_0)$. As x_1, x_2, u_1 and u_2 are basic variables in the optimal we have,

$$B_{\mathcal{J}_0}(s) = B + \begin{pmatrix} I_4 \\ 0 \end{pmatrix} \begin{pmatrix} s - s_0 & 0 & 0 & 0 \\ 0 & \frac{s_0 - s}{s_0 s} & 0 & 0 \\ 0 & 0 & s_0 - s & 0 \\ 0 & 0 & 0 & \frac{s - s_0}{s_0 s} \end{pmatrix} (I_4 \quad 0) =: B + XSY.$$

Therefore, $det(B_{\mathcal{J}_0}(s)) = det[B(I + B^{-1}XSY)] = det(B)det(I + B^{-1}XSY) = det(B)det(I_4 + SYB^{-1}X)$. Note that $\alpha(s) = det(I_4 + SYB^{-1}X)$ and therefore, it is clear that the inverse of $B_{\mathcal{J}_0}(s)$ exists if and only if $\alpha(s) \neq 0$. Assuming $\alpha(s) \neq 0$ we find an expression for $B^{-1}_{\mathcal{J}_0}(s)$ as follows:

$$\begin{aligned} B^{-1}_{\mathcal{J}_0}(s) &= B^{-1}(I + XSYB^{-1})^{-1} \\ &= B^{-1}[I - (I + XSYB^{-1})^{-1} XSYB^{-1}] \\ &= B^{-1} - B^{-1}X(I_4 + SYB^{-1}X)^{-1} SYB^{-1}. \end{aligned}$$

Now,

$$\begin{aligned} \mathbf{x}_{\mathcal{J}_0}(s) &= B^{-1}_{\mathcal{J}_0}(s)\mathbf{b}(s) \\ &= [B^{-1} - B^{-1}X(I_4 + SYB^{-1}X)^{-1} SYB^{-1}]\mathbf{b}(s) \\ &= B^{-1}\mathbf{b}(s) - B^{-1}X(I_4 + SYB^{-1}X)^{-1} SYB^{-1}\mathbf{b}(s) \\ &= B^{-1}\mathbf{b}(s) - \left[\beta^1\ \beta^2\ \beta^3\ \beta^4\right](I_4 + SYB^{-1}X)^{-1} S \begin{bmatrix} (x_B(s))_1 \\ (x_B(s))_2 \\ (x_B(s))_3 \\ (x_B(s))_4 \end{bmatrix} \\ &= x_B(s) - \left[\beta^1\ \beta^2\ \beta^3\ \beta^4\right] R(s) \begin{bmatrix} (x_B(s))_1 \\ (x_B(s))_2 \\ (x_B(s))_3 \\ (x_B(s))_4 \end{bmatrix}, \end{aligned}$$

where we have defined $R(s) := (I_4 + SYB^{-1}X)^{-1}S$ and $x_B(s) := B^{-1}\mathbf{b}(s)$. An expression for the 4×4 matrix $R(s)$ can be found easily. Note that if $\alpha(s) = 0$ then $B_{\mathcal{J}_0}(s)$ is not invertible and by definition it follows that $\mathbf{x}_{\mathcal{J}_0}(s) = 0$. □

Definition 9.4.2. *Given $s_0 > 0$. Let $\mathcal{J}_0$ be an optimal-basis-index for the problem $(QP(s_0))$. Define*

$$Reg(\mathcal{J}_0) := \{s > 0 : \alpha(s) \neq 0, (\mathbf{x}_{\mathcal{J}_0}(s))_i \geq 0 \text{ for all } i = 1, \ldots, m\}.$$

Note that $\mathbf{x}_{\mathcal{J}_0}(s)$ is a rational function of s and therefore $Reg(\mathcal{J}_0)$ is a union of closed intervals except for the roots of $\alpha(s) = 0$. Determining $Reg(\mathcal{J}_0)$ is therefore an easy task.

Theorem 9.4.2. *Let $s_0 > 0$ and let $\mathcal{J}_0$ be an optimal-basis-index with x_B as the basic feasible solution for the problem $(QP(s_0))$. Suppose u_1 and u_2*

are basic variables in the optimal solution. Then $B_{\mathcal{J}_0}(s)$ is an optimal basis for $(QP(s))$ if and only if s in $Reg(\mathcal{J}_0)$. Suppose, s in $Reg(\mathcal{J}_0)$ then the objective value of $(QP(s))$ is given by

$$\gamma_{\mathcal{J}_0}(s) = c_B^T(s)x_B(s) - c_B^T(s)\ \left[\beta^1\ \beta^2\ \beta^3\ \beta^4\right] R(s) \begin{bmatrix} (x_B(s))_1 \\ (x_B(s))_2 \\ (x_B(s))_3 \\ (x_B(s))_4 \end{bmatrix}.$$

Proof. Suppose s in $Reg(\mathcal{J}_0)$. Then, $B_{\mathcal{J}_0}(s)$ has linearly independent columns (because $\alpha(s) \neq 0$). As $\mathbf{x}_{\mathcal{J}_0}(s) := B_{\mathcal{J}_0}^{-1}(s)\mathbf{b}(s)$ we know that $\mathbf{x}_{\mathcal{J}_0}(s)$ is a basic solution. $\mathbf{x}_{\mathcal{J}_0}(s)$ is a feasible solution because

$$(\mathbf{x}_{\mathcal{J}_0}(s))_i \geq 0 \text{ for all } i = 1, \ldots, m.$$

If $s > 0$ is such that $s \notin Reg(\mathcal{J}_0)$ then either feasibility is lost or the columns of $B_{\mathcal{J}_0}(s)$ are not independent. This proves the first part of the theorem.

If the solution is optimal for $(QP(s))$ then the optimal objective value is given by

$$\begin{aligned}\gamma_{\mathcal{J}_0}(s) &= c_B^T(s)\mathbf{x}_{\mathcal{J}}(s) \\ &= c_B^T(s)\{x_B(s) - \left[\beta^1\ \beta^2\ \beta^3\ \beta^4\right] R(s) \begin{bmatrix} (x_B(s))_1 \\ (x_B(s))_2 \\ (x_B(s))_3 \\ (x_B(s))_4 \end{bmatrix}\} \\ &= c_B^T(s)x_B(s) - c_B^T(s)\ \left[\beta^1\ \beta^2\ \beta^3\ \beta^4\right] R(s) \begin{bmatrix} (x_B(s))_1 \\ (x_B(s))_2 \\ (x_B(s))_3 \\ (x_B(s))_4 \end{bmatrix}.\end{aligned}$$

This proves the theoren. □

We now present the following theorem which gives a way to compute γ_{opt}.

Theorem 9.4.3. *There exists a finite set of basis indices $\mathcal{J}_0, \mathcal{J}_1, \ldots, \mathcal{J}_l$ such that $R^+ = \cup_{k=1}^{l} Reg(\mathcal{J}_k)$. Furthermore if*

$$f_k := \min_{s \in Reg(\mathcal{J}_k)} \gamma_{\mathcal{J}_k}(s)$$

then

$$\gamma_{opt} = \min_{k=0,\ldots,l} f_k.$$

Proof. The proof is iterative:

Step 1) Let $s_1 > 0$. Find $\mathcal{J}_1, Reg(\mathcal{J}_1)$ and f_1 where $\mathcal{J}_1$ is the optimal basis-index for $(QP(s_1))$. Note that $Reg(\mathcal{J}_1)$ is a finite union of closed intervals except for a finite number of points which can be determined.

Step 2) Suppose we have reached the $(k-1)^{th}$ step. If $\cup_{p=1}^{k-1} Reg(\mathcal{J}_p) = R^+$ then stop and the theorem is true with $l = k - 1$. Otherwise choose any s in $R^+ - \cup_{p=1}^{k-1} Reg(\mathcal{J}_p)$ and perform step 1 with $s_1 = s$.

This procedure has to terminate because for any s in R^+ there exists a basic feasible solution and there are only finite number of basis-index sets.

□

We have assumed that for $(QP(s_0))$ the optimal is such that u_1, u_2 are there in the basis (we assume that x_1 and x_2 are always in the optimal basis). This might not be so. In that case the expressions can be easily modified and they will be simpler than the ones derived.

9.5 Summary

A problem which incorporates $\mathcal{H}_2$ nominal performance and ℓ_1 robust performance was formulated. It was shown that this problem can be solved via quadratic programming using sensitivity techniques.

References

1. W. Rudin. *Principles of Mathematical Analysis.* McGraw-Hill, Inc., 1976.
2. C. Chen. *Linear System Theory And Design.* Holt, Rinehart And Winston, Inc., New York, 1984.
3. M. A. Dahleh and I. J. Diaz-Bobillo. *Control of Uncertain Systems: A Linear Programming Approach.* Prentice Hall, Englewood Cliffs, New Jersey, 1995.
4. J. C. Doyle, K. Glover, P. Khargonekar, and B. A. Francis. State space solutions to standard $\mathcal{H}_2$ and $\mathcal{H}_\infty$ control problems. *IEEE Trans. Automat. Control*, 34, no. 8:pp. 831–847, 1989.
5. M. A. Dahleh and J. B. Pearson. ℓ_1 Optimal feedback controllers for MIMO discrete-time systems. *IEEE Trans. Automat. Control*, 32, no. 4:pp. 314–322, 1987.
6. J. C. Doyle, K. Zhou, and B. Bodenheimer. Optimal control with mixed $\mathcal{H}_2$ and $\mathcal{H}_\infty$ performance objectives. In *Proceedings of the American Control Conference.* Vol. 3, pp. 2065-2070, Pittsburg, PA, June 1989.
7. P. P. Khargaonekar and M. A. Rotea. Mixed $\mathcal{H}_2/\mathcal{H}_\infty$ control; a convex optimization approach. *IEEE Trans. Automat. Control*, 36, no. 7:pp. 824–837, 1991.
8. N. Elia, M. A. Dahleh, and I. J. Diaz-Bobillo. Controller design via infinite dimensional linear programming. In *Proceedings of the American Control Conference.* Vol. 3, pp. 2165-2169, San Fransisco, California, June 1993.
9. M. Sznaier. Mixed $\ell_1/\mathcal{H}_\infty$ controllers for MIMO discrete time systems. In *Proceedings of the IEEE Conference on Decision and Control.* pp. 3187-3191, Orlando, Florida, December 1994.
10. H. Rotstein and A. Sideris. $\mathcal{H}_\infty$ optimization with time domain constarints. *IEEE Trans. Automat. Control*, 39:pp. 762–770, 1994.
11. X. Chen and J. Wen. A linear matrix inequality approach to discrete-time $\ell_1/\mathcal{H}_\infty$ control problems. In *Proceedings of the IEEE Conference on Decision and Control.* pp: 3670-3675, New Orleans, LA, December 1995.
12. N. Elia and M. A. Dahleh. ℓ_1 minimization with magnitude constraints in the frequency domain. *Journal of Optimization Theory and its Applications*, 93:27–52, 1997.
13. N. Elia, P. M. Young, and M. A. Dahleh. Multiobjective control via infinite dimensional lmi optimization. In *Proceedings of the Allerton Conference on Communication, Control and Computing.* pp: 186-195, Urbana, Il., 1995.
14. P. Voulgaris. Optimal $\mathcal{H}_2/\ell_1$ control via duality theory. *IEEE Trans. Automat. Control*, 4, no. 11:pp. 1881–1888, 1995.
15. M. V. Salapaka, M. Dahleh, and P. Voulgaris. Mixed objective control synthesis: Optimal $\ell_1/\mathcal{H}_2$ control. *SIAM Journal on Control and Optimization*, V35 N5:1672–1689, 1997.

16. M. V. Salapaka, P. Voulgaris, and M. Dahleh. SISO controller design to minimize a positive combination of the ℓ_1 and the $\mathcal{H}_2$ norms. *Automatica*, 33 no. 3:387–391, 1997.
17. M. V. Salapaka, P. Voulgaris, and M. Dahleh. Controller design to optimize a composite performance measure. *Journal of Optimization Theory and its Applications*, 91 no. 1:91–113, 1996.
18. M. V. Salapaka, M. Khammash, and M. Dahleh. Solution of mimo $\mathcal{H}_2/\ \ell_1$ problem without zero interpolation. In *Proceedings of the IEEE Conference on Decision and Control.* pp: 1546-1551, San Diego, CA, December 1997.
19. P. M. Young and M. A. Dahleh. Infinite dimensional convex optimization in optimal and robust control. *IEEE Trans. Automat. Control*, 12, 1997.
20. N. Elia and M. A. Dahleh. Controller design with multiple objectives. *IEEE Trans. Automat. Control*, 42, no. 5:596–613, 1997.
21. M. V. Salapaka, M. Dahleh, and P. Voulgaris. Mimo optimal control design: the interplay of the $\mathcal{H}_2$ and the ℓ_1 norms. *IEEE Trans. Automat. Control*, 43, no. 10:1374–1388, 1998.
22. S. P. Boyd and C. H. Barratt. *Linear Controller Design: Limits Of Performance.* Prentice Hall, Englewood Cliffs, New Jersey, 1991.
23. N. O. D. Cuhna and E. Polak. Constrained minimization under vector valued criteria in finite dimensional spaces. *J. Math. Annal. and Appl.*, 19:pp 103–124, 1967.
24. M. Khammash. Solution of the ℓ_1 mimo control problem without zero interpolation. In *Proceedings of the IEEE Conference on Decision and Control.* pp: 4040-4045, Kobe, Japan, December 1996.
25. J. C. Doyle. Analysis of feedback systems with structured uncertainty. In *IEE Proceedings.* Vol. 129-D(6), pp. 242-250, November 1982.
26. M. H. Khammash and J. B. Pearson. Robust disturbance rejection in ℓ_1-optimal control systems. *Systems and Control letters*, 14, no. 2:pp. 93–101, 1990.
27. M. H. Khammash and J. B. Pearson. Performance robustness of discrete-time systems with structured uncertainty. *IEEE Trans. Automat. Control*, 36, no. 4:pp. 398–412, 1991.
28. M. H. Khammash and J. B. Pearson. Analysis and design for robust performance with structured uncertainty. *Systems and Control letters*, 20, no. 3:pp. 179–187, 1993.
29. M. H. Khammash. Synthesis of globally optimal controllers for robust performance to unstructured uncertainty. *IEEE Trans. Automat. Control*, 41:189–198, 1996.

Index

Lecture Notes in Control and Information Sciences

Edited by M. Thoma

1993–1999 Published Titles:

Vol. 186: Sreenath, N.
Systems Representation of Global Climate Change Models. Foundation for a Systems Science Approach.
288 pp. 1993 [3-540-19824-5]

Vol. 187: Morecki, A.; Bianchi, G.; Jaworeck, K. (Eds)
RoManSy 9: Proceedings of the Ninth CISM-IFToMM Symposium on Theory and Practice of Robots and Manipulators.
476 pp. 1993 [3-540-19834-2]

Vol. 188: Naidu, D. Subbaram
Aeroassisted Orbital Transfer: Guidance and Control Strategies
192 pp. 1993 [3-540-19819-9]

Vol. 189: Ilchmann, A.
Non-Identifier-Based High-Gain Adaptive Control
220 pp. 1993 [3-540-19845-8]

Vol. 190: Chatila, R.; Hirzinger, G. (Eds)
Experimental Robotics II: The 2nd International Symposium, Toulouse, France, June 25-27 1991
580 pp. 1993 [3-540-19851-2]

Vol. 191: Blondel, V.
Simultaneous Stabilization of Linear Systems
212 pp. 1993 [3-540-19862-8]

Vol. 192: Smith, R.S.; Dahleh, M. (Eds)
The Modeling of Uncertainty in Control Systems
412 pp. 1993 [3-540-19870-9]

Vol. 193: Zinober, A.S.I. (Ed.)
Variable Structure and Lyapunov Control
428 pp. 1993 [3-540-19869-5]

Vol. 194: Cao, Xi-Ren
Realization Probabilities: The Dynamics of Queuing Systems
336 pp. 1993 [3-540-19872-5]

Vol. 195: Liu, D.; Michel, A.N.
Dynamical Systems with Saturation Nonlinearities: Analysis and Design
212 pp. 1994 [3-540-19888-1]

Vol. 196: Battilotti, S.
Noninteracting Control with Stability for Nonlinear Systems
196 pp. 1994 [3-540-19891-1]

Vol. 197: Henry, J.; Yvon, J.P. (Eds)
System Modelling and Optimization
975 pp approx. 1994 [3-540-19893-8]

Vol. 198: Winter, H.; Nüßer, H.-G. (Eds)
Advanced Technologies for Air Traffic Flow Management
225 pp approx. 1994 [3-540-19895-4]

Vol. 199: Cohen, G.; Quadrat, J.-P. (Eds)
11th International Conference on Analysis and Optimization of Systems – Discrete Event Systems: Sophia-Antipolis, June 15–16–17, 1994
648 pp. 1994 [3-540-19896-2]

Vol. 200: Yoshikawa, T.; Miyazaki, F. (Eds)
Experimental Robotics III: The 3rd International Symposium, Kyoto, Japan, October 28-30, 1993
624 pp. 1994 [3-540-19905-5]

Vol. 201: Kogan, J.
Robust Stability and Convexity
192 pp. 1994 [3-540-19919-5]

Vol. 202: Francis, B.A.; Tannenbaum, A.R. (Eds)
Feedback Control, Nonlinear Systems, and Complexity
288 pp. 1995 [3-540-19943-8]

Vol. 203: Popkov, Y.S.
Macrosystems Theory and its Applications: Equilibrium Models
344 pp. 1995 [3-540-19955-1]

Vol. 204: Takahashi, S.; Takahara, Y.
Logical Approach to Systems Theory
192 pp. 1995 [3-540-19956-X]

Vol. 205: Kotta, U.
Inversion Method in the Discrete-time Nonlinear Control Systems Synthesis Problems
168 pp. 1995 [3-540-19966-7]

Vol. 206: Aganovic, Z.; Gajic, Z.
Linear Optimal Control of Bilinear Systems with Applications to Singular Perturbations and Weak Coupling
133 pp. 1995 [3-540-19976-4]

Vol. 207: Gabasov, R.; Kirillova, F.M.; Prischepova, S.V.
Optimal Feedback Control
224 pp. 1995 [3-540-19991-8]

Vol. 208: Khalil, H.K.; Chow, J.H.; Ioannou, P.A. (Eds)
Proceedings of Workshop on Advances inControl and its Applications
300 pp. 1995 [3-540-19993-4]

Vol. 209: Foias, C.; Özbay, H.; Tannenbaum, A.
Robust Control of Infinite Dimensional Systems: Frequency Domain Methods
230 pp. 1995 [3-540-19994-2]

Vol. 210: De Wilde, P.
Neural Network Models: An Analysis
164 pp. 1996 [3-540-19995-0]

Vol. 211: Gawronski, W.
Balanced Control of Flexible Structures
280 pp. 1996 [3-540-76017-2]

Vol. 212: Sanchez, A.
Formal Specification and Synthesis of Procedural Controllers for Process Systems
248 pp. 1996 [3-540-76021-0]

Vol. 213: Patra, A.; Rao, G.P.
General Hybrid Orthogonal Functions and their Applications in Systems and Control
144 pp. 1996 [3-540-76039-3]

Vol. 214: Yin, G.; Zhang, Q. (Eds)
Recent Advances in Control and Optimization of Manufacturing Systems
240 pp. 1996 [3-540-76055-5]

Vol. 215: Bonivento, C.; Marro, G.; Zanasi, R. (Eds)
Colloquium on Automatic Control
240 pp. 1996 [3-540-76060-1]

Vol. 216: Kulhavý, R.
Recursive Nonlinear Estimation: A Geometric Approach
244 pp. 1996 [3-540-76063-6]

Vol. 217: Garofalo, F.; Glielmo, L. (Eds)
Robust Control via Variable Structure and Lyapunov Techniques
336 pp. 1996 [3-540-76067-9]

Vol. 218: van der Schaft, A.
L_2 Gain and Passivity Techniques in Nonlinear Control
176 pp. 1996 [3-540-76074-1]

Vol. 219: Berger, M.-O.; Deriche, R.; Herlin, I.; Jaffré, J.; Morel, J.-M. (Eds)
ICAOS '96: 12th International Conference on Analysis and Optimization of Systems - Images, Wavelets and PDEs:
Paris, June 26-28 1996
378 pp. 1996 [3-540-76076-8]

Vol. 220: Brogliato, B.
Nonsmooth Impact Mechanics: Models, Dynamics and Control
420 pp. 1996 [3-540-76079-2]

Vol. 221: Kelkar, A.; Joshi, S.
Control of Nonlinear Multibody Flexible Space Structures
160 pp. 1996 [3-540-76093-8]

Vol. 222: Morse, A.S.
Control Using Logic-Based Switching
288 pp. 1997 [3-540-76097-0]

Vol. 223: Khatib, O.; Salisbury, J.K.
Experimental Robotics IV: The 4th International Symposium, Stanford, California, June 30 - July 2, 1995
596 pp. 1997 [3-540-76133-0]

Vol. 224: Magni, J.-F.; Bennani, S.; Terlouw, J. (Eds)
Robust Flight Control: A Design Challenge
664 pp. 1997 [3-540-76151-9]

Vol. 225: Poznyak, A.S.; Najim, K.
Learning Automata and Stochastic Optimization
219 pp. 1997 [3-540-76154-3]

Vol. 226: Cooperman, G.; Michler, G.; Vinck, H. (Eds)
Workshop on High Performance Computing and Gigabit Local Area Networks
248 pp. 1997 [3-540-76169-1]

Vol. 227: Tarbouriech, S.; Garcia, G. (Eds)
Control of Uncertain Systems with Bounded Inputs
203 pp. 1997 [3-540-76183-7]

Vol. 228: Dugard, L.; Verriest, E.I. (Eds)
Stability and Control of Time-delay Systems
344 pp. 1998 [3-540-76193-4]

Vol. 229: Laumond, J.-P. (Ed.)
Robot Motion Planning and Control
360 pp. 1998 [3-540-76219-1]

Vol. 230: Siciliano, B.; Valavanis, K.P. (Eds)
Control Problems in Robotics and Automation
328 pp. 1998 [3-540-76220-5]

Vol. 231: Emel'yanov, S.V.; Burovoi, I.A.; Levada, F.Yu.
Control of Indefinite Nonlinear Dynamic Systems
196 pp. 1998 [3-540-76245-0]

Vol. 232: Casals, A.; de Almeida, A.T. (Eds)
Experimental Robotics V: The Fifth International Symposium Barcelona, Catalonia, June 15-18, 1997
190 pp. 1998 [3-540-76218-3]

Vol. 233: Chiacchio, P.; Chiaverini, S. (Eds)
Complex Robotic Systems
189 pp. 1998 [3-540-76265-5]

Vol. 234: Arena, P.; Fortuna, L.; Muscato, G.; Xibilia, M.G.
Neural Networks in Multidimensional Domains: Fundamentals and New Trends in Modelling and Control
179 pp. 1998 [1-85233-006-6]

Vol. 235: Chen, B.M.
H∞ Control and Its Applications
361 pp. 1998 [1-85233-026-0]

Vol. 236: de Almeida, A.T.; Khatib, O. (Eds)
Autonomous Robotic Systems
283 pp. 1998 [1-85233-036-8]

Vol. 237: Kreigman, D.J.; Hagar, G.D.; Morse, A.S. (Eds)
The Confluence of Vision and Control
304 pp. 1998 [1-85233-025-2]

Vol. 238: Elia, N. ; Dahleh, M.A.
Computational Methods for Controller Design
200 pp. 1998 [1-85233-075-9]

Vol. 239: Wang, Q.G.; Lee, T.H.; Tan, K.K.
Finite Spectrum Assignment for Time-Delay Systems
200 pp. 1998 [1-85233-065-1]

Vol. 240: Lin, Z.
Low Gain Feedback
376 pp. 1999 [1-85233-081-3]

Vol. 241: Yamamoto, Y.; Hara S.
Learning, Control and Hybrid Systems
472 pp. 1999 [1-85233-076-7]

Vol. 242: Conte, G.; Moog, C.H.; Perdon A.M.
Nonlinear Control Systems
192 pp. 1999 [1-85233-151-8]

Vol. 243: Tzafestas, S.G.; Schmidt, G. (Eds)
Progress in Systems and Robot Analysis and Control Design
624 pp. 1999 [1-85233-123-2]

Vol. 244: Nijmeijer, H.; Fossen, T.I. (Eds)
New Directions in Nonlinear Observer Design
552pp: 1999 [1-85233-134-8]

Vol. 245: Garulli, A.; Tesi, A.; Vicino, A. (Eds)
Robustness in Identification and Control
448pp: 1999 [1-85233-179-8]

Vol. 246: Aeyels, D.;
Lamnabhi-Lagarrigue, F.; van der Schaft, A. (Eds)
Stability and Stabilization of Nonlinear Systems
408pp: 1999 [1-85233-638-2]

Vol. 247: Young, K.D.; Özgüner, Ü. (Eds)
Variable Structure Systems, Sliding Mode and Nonlinear Control
400pp: 1999 [1-85233-197-6]

Vol. 248: Chen, Y.; Wen C.
Iterative Learning Control
216pp: 1999 [1-85233-190-9]

Vol. 249: Cooperman, G.; Jessen, E.;
Michler, G. (Eds)
Workshop on Wide Area Networks and High Performance Computing
352pp: 1999 [1-85233-642-0]

Vol. 250: Corke, P. ; Trevelyan, J. (Eds)
Experimental Robotics VI
552pp: 2000 [1-85233-210-7]

Vol. 251: van der Schaft, A. ; Schumacher, J.
An Introduction to Hybrid Dynamical Systems
192pp: 2000 [1-85233-233-6]